Applied Science Review™

Zoology

Applied Science Review™

Zoology

Mimi Bres, PhD
Assistant Professor of Biology
Philadelphia College of Pharmacy and Science
Philadelphia

Springhouse Corporation
Springhouse, Pennsylvania

Staff

EXECUTIVE DIRECTOR, EDITORIAL
Stanley Loeb

PUBLISHER, TRADE AND TEXTBOOKS
Minnie B. Rose, RN, BSN, MEd

ART DIRECTOR
John Hubbard

CLINICAL CONSULTANT
Maryann Foley, RN, BSN

EDITORS
Diane Labus, David Moreau, Karen Zimmerman

COPY EDITORS
Diane M. Armento, Pamela Wingrod

DESIGNERS
Stephanie Peters (associate art director), Matie Patterson (senior designer)

COVER ILLUSTRATION
Scott Thorn Barrows

ILLUSTRATORS
Jean Gardner, Robert Neumann, Judy Newhouse

MANUFACTURING
Deborah Meiris (director), Anna Brindisi, Kate Davis, T.A. Landis

EDITORIAL ASSISTANTS
Caroline Lemoine, Louise Quinn, Betsy K. Snyder

Printed in the United States of America.
ASR7-010793

Library of Congress Cataloging-in-Publication Data
Bres, Mimi.
Zoology / Mimi Bres.
p. cm.–(Applied science review)
Includes bibliographical references and index
1. Zoology–Outlines, syllabi, etc. I. Title II. Series.
QL52.B73 1994
591–dc20 93-1659
ISBN 0-87434-571-5 CIP

Contents

Advisory Board

Reviewers

Dedication

To all the professors who have helped me discover the
value of teaching and learning.
MB

Preface

This book is one in a series designed to help students learn and study scientific concepts and essential information covered in core science subjects. Each book offers a comprehensive overview of a scientific subject as taught at the college or university level and features numerous illustrations and charts to enhance learning and studying. Each chapter includes a list of objectives, a detailed outline covering a course topic, and assorted study activities. A glossary appears at the end of each book; terms that appear in the glossary are highlighted throughout the book in boldface italic type.

Zoology, the study of animals, provides conceptual and factual information on the various topics covered in most zoology courses and textbooks and focuses on helping students to understand:

- the evolution and diversification of animals
- the classification and phylogeny of animals
- similarities and differences in animal body plans
- identifying features and characteristics of various animal phyla
- the ecologic relationships among different animal phyla
- principles of animal behavior
- the significance of animal ecology.

1

Overview of Zoology

Objectives

After studying this chapter, the reader should be able to:
• Identify and characterize various branches of zoology.
• Understand the basic principles of the scientific method.
• Differentiate between a hypothesis and a theory.

I. The Science of Zoology

A. General information

1. *Zoology*, the scientific study of animals, is a subdivision of *biology*, which is the study of living organisms
2. Zoology can be subdivided into various branches (specialized areas of study)
    a. *Entomology* is the study of insects
    b. *Mammalogy* is the study of mammals
    c. *Ornithology* is the study of birds
    d. *Ichthyology* is the study of fish
    e. *Herpetology* is the study of reptiles and amphibians
    f. *Embryology* is the study of growth and development of animal embryos
3. Within each branch, zoologists specialize further
    a. Some zoologists concentrate on specific anatomic or physiologic characteristics; others study animal behavior or ecology
    b. Zoologists also may specialize in a specific species or group of species
4. Zoology is a comparative science; descriptions of anatomic and physiologic characteristics of one species are more meaningful and informative when seen in relation to other species in the animal kingdom

B. Animal diversity

1. Most of the organisms alive today belong to one kingdom—the Kingdom Animalia, which encompasses all multicellular animals (see Chapter 4, Classification and Phylogeny of Animals)
2. Scientists have identified and described more than 1 million living species of animals
3. The current rate at which new species are being identified leads scientists to estimate that millions more remain undiscovered
4. This extraordinary array of animals is the outcome of hundreds of millions of years of evolution

a. Evidence that species evolve comes from lines of investigation that began nearly two centuries ago
b. Opposition to ideas of evolutionary descent were put forward by theologians, who interpreted the Old Testament account of creation literally
c. Although controversy about evolution still exists, overwhelming evidence from a variety of sources substantiates the occurrence of evolutionary change (see Chapter 2, Evolution of Animal Diversity)

5. All living organisms share common characteristics and a common ancestry; modern animals are the descendants of successful lineages

II. Basic Principles of the Scientific Method

A. General information

1. There are many different methods for approaching and solving problems
2. The process known as the scientific method is a series of steps, or guidelines, often followed by scientists to answer questions about nature
3. Zoologists use the scientific method to investigate questions about the anatomy, physiology, behavior, and ecology of animals

B. Steps of the scientific method

1. Although there is no single scientific method for investigating zoologic questions, the sequence of activities listed below provides a rough outline of how a scientist might approach such an investigation
2. The scientific method begins with an observation or question about some aspect of the natural world
3. To approach this question, scientists develop one or more *hypotheses* (provisional or tentative answers to the question under investigation)
4. Hypotheses are tested by experimentation or additional observations
   a. To test a hypothesis, a scientist makes predictions about what will be observed if the hypothesis is correct
   b. These predictions take the form of an "if-then" statement; if my hypothesis is true, then I will expect to observe the following events
   c. Testing hypotheses involves analyzing observations or experimental results to determine whether they match the predicted (expected) observations
5. The process is concluded with a statement about whether the original hypothesis can be accepted or rejected, based on the results of the tested predictions

C. Differentiating between hypotheses and theories

1. A scientific hypothesis is a tentative explanation for some event (an educated guess)
2. A scientific *theory* is a system of ideas and concepts assembled to organize and explain related events or observations
   a. A theory is composed of a related set of hypotheses
   b. When these hypotheses are combined, they provide a broad explanation for some fundamental aspect of the natural world
3. A theory differs from a hypothesis in its breadth or scope of application
4. Like hypotheses, theories are tested, revised, and provisionally accepted or rejected; theories are accepted by the scientific community only when they have been verified by a large number of repeated observations and tests

5. Darwin's theory of natural selection as a mechanism of evolution is a good example; this theory provides a broad explanation for a large group of observable phenomena and is supported by a comprehensive array of evidence

Study Activities

1. Write a brief essay that explains why an organized series of steps, such as those of the scientific method, would be helpful in zoologic investigations.
2. Compare the terms theory and hypothesis and give an example of each.

2

Evolution of Animal Diversity

Objectives

After studying this chapter, the reader should be able to:
- Understand the basic concepts of evolution.
- Document the history of evolutionary thought.
- Understand and apply Darwin's theory of natural selection.
- Distinguish between the modern synthesis and punctuated equilibrium.
- Identify several lines of evidence that support the theory of evolution.
- Compare the mechanisms of microevolution with those of macroevolution and explain how each affects speciation.

I. Basic Concepts of Evolution

A. General information

1. The theory of evolution is the underlying thread that ties the science of biology together
2. Evolution explains four basic observations about living organisms
    a. All living organisms share basic characteristics, thereby demonstrating unity of life
        (1) All living organisms use nucleic acids—deoxyribonucleic acid (DNA) or ribonucleic acid (RNA)—for transmitting hereditary information; the genetic code is the same for all living organisms
        (2) All living organisms are composed of cells; some are unicellular, whereas others are multicellular
        (3) All living organisms use the same 20 amino acids to synthesize proteins
        (4) All ***aerobic*** (oxygen-using) organisms use the molecule ***adenosine triphosphate (ATP)*** for energy storage
    b. Living organisms are incredibly diverse
        (1) To date, about 2 million species have been named and described
        (2) Biologists estimate that between 4 and 30 million unidentified species have yet to be discovered
    c. Living organisms are well adapted to their environments; birds, for example, have several adaptations that facilitate flight
        (1) Bird feathers are light and strong
        (2) Bird bones are thin-walled and light, which decreases body weight

(3) Birds have unique, efficient respiratory systems that provide the extra oxygen needed for flight

d. Living organisms share certain biologic patterns, which are evident from the history of life on earth

(1) All living organisms show evidence of a common ancestry

(2) The presence of ***homologous*** characteristics among organisms physically supports the notion of a common ancestry

(3) Mammalian forelimbs, which are all constructed from the same skeletal elements, are examples of homologous structures

(4) Vestigial organs, historical remnants of structures that had important functions in ancestors but no longer are essential, also are homologous; these include the pelvis and leg bones of snakes and whales, as well as the human appendix

B. History of evolutionary thought

1. The first convincing case for evolution was presented by Charles Darwin in 1859
2. Pre-Darwinian views about the nature of life generally did not include evolution
3. The most influential Greek philosophers, Aristotle and Plato, did not accept the idea of evolution

a. However, Aristotle did develop a hierarchy of increasing complexity upon which he placed all living organisms

b. He named this hierarchy the *scala naturae* (scale of nature)

c. He viewed species as fixed and immutable; each species had a fixed place on the scala naturae that would never change

4. European naturalists of the 18th and 19th centuries also viewed species as immutable

a. This view originated in the Old Testament account of creation

b. Naturalists of that era viewed the adaptations of organisms as evidence that the Creator had designed every species for a particular purpose

c. Georges Cuvier, a French anatomist, founded the science of ***paleontology*** (the study of fossils) in the early 1800s; he used evidence from the fossil record to oppose the idea of evolution

(1) Cuvier discovered that many rock strata (layers) had assemblages of characteristic fossils that were not found in the layers above and below that stratum

(2) He also recognized that the history of life was filled with extinctions of species and the development of new species, but he did not wish to attribute these changes to evolution

(3) Cuvier developed the theory of ***catastrophism*** to explain his observations

(a) According to this theory, boundaries between fossil layers correspond to environmental catastrophes (such as floods) in the geologic past

(b) Since there were many rock strata, there have been many catastrophic events in the earth's past

(c) Subsequent to each catastrophe, plants and animals were repopulated by migration from unaffected areas or by new creation events

(d) Cuvier concluded that the succession of species in the fossil record was due not to evolution but to new populations of plants and animals that had migrated into the area

d. Jean Baptiste Lamarck, a contemporary of Cuvier, was an influential French invertebrate zoologist; he proposed a theory on the ***inheritance of acquired characteristics*** to account for the evolution of animal adaptations

(1) According to his theory, organisms accumulate adaptations by constantly striving to adjust to their environment

(2) These adaptations are passed on by heredity to the offspring; for example, the long neck of the giraffe evolved gradually as a result of several generations of ancestors reaching higher and higher into the tree tops for food

(3) At present, no evidence supports the theory of inheritance of acquired characteristics in multicellular animals

II. Darwin's Theory of Natural Selection

A. General information

1. The theory of ***natural selection,*** as presented by Charles Darwin in his book *The Origin of Species* in 1859, forms the basis of modern ideas about animal adaptation
2. Darwin viewed the process of evolution as descent with modification
   a. According to this theory, all living organisms are descended from a common ancestor that lived in the remote past
   b. During many generations of descent, animals accumulated a variety of modifications (adaptations) that enabled them to live in diverse habitats
   c. Darwin viewed the history of life as a branching tree (his analogy for descent with modification)
      (1) Multiple branches extend out from the main trunk
      (2) The branches keep forking and dividing, all the way down to tiny twigs
      (3) At each fork, there is an ancestor common to all lines of evolution branching from that fork
      (4) Species that are closely related, such as the wolf and domestic dog, have similar characteristics because they branched from each other (diverged) at the same twig level on the evolutionary tree

B. Theory highlights

1. Darwin's theory of natural selection can be viewed as a series of logical deductions; the theory is based on observable facts and supported by overwhelming evidence
2. All organisms show variation
   a. Darwin and others of his generation were very familiar with variation among domestic plants and animals
   b. Artificial selection (controlled breeding) had been practiced for thousands of years
3. Much of this observable variation is heritable (passed on through genetics)
4. All organisms produce more offspring than can survive (for example, one female fish may lay hundreds of eggs, only a fraction of which will survive)

5. Since natural resources are limited, not all offspring survive to become adults and reproduce (although each female fish lays hundreds of eggs, the sea does not fill up with fish)
   a. Offspring must struggle for the scarce natural resources, such as food and shelter
   b. Survival also is a measure of the offspring's physical stamina and resistance to adverse environmental conditions, such as bad weather or disease
6. Because of natural variation, some individuals are born with a more favorable set of characteristics than others
   a. Favorable characteristics are those that best fit the individual to its environment
   b. Individuals with these favorable characteristics are more likely to survive and reproduce, passing these characteristics on to their offspring
   c. Gradually, over generations, the percentage of the population that possesses these favorable traits increases
   d. This change in the frequency (percentage) of a trait in the population is called ***evolution***
7. When enough differences accumulate in a population to prevent interbreeding with the ancestral population, the new population is considered a separate, new species; thus, new species originate through natural selection

C. Modern evolutionary theory

1. Evolution does not depend on any single mechanism (such as natural selection, genetic drift, or mutations); available evidence suggests several possible mechanisms of evolutionary change that are not mutually exclusive
2. At present, there are two prevailing ideas about the ***tempo*** (rate) and ***mode*** (mechanisms) of evolution
   a. The *modern synthesis* was proposed in the 1940s, after the genetic basis of heredity and natural selection had been confirmed; the synthesis was forged by scientists from various branches of biology, geology, and geography
      (1) The modern synthesis holds that evolutionary change results primarily from the action of natural selection on variation within populations
      (2) It focuses on genes and changes in gene frequencies within populations
      (3) New species arise from the gradual accumulation of adaptive characteristics within a population
   b. Another theory, ***punctuated equilibrium,*** proposed by Niles Eldredge and Steven Jay Gould in 1972, suggests that evolutionary change occurs in spurts
      (1) According to their theory, periods of active evolution are relatively short in terms of geologic time
      (2) A short period of rapid evolution will be followed by a long period when no change occurs (called *stasis*)
      (3) ***Speciation*** (the development of new species) may be triggered by major environmental changes or major genetic changes

III. Evidence for Evolution

A. General information

1. Direct evidence for evolution can be obtained from several fields of study, including ecology, paleontology, anatomy, and physiology
2. Evidence such as the fossil record, classification systems, geographic distribution of species, embryonic development, and molecular biology studies all support the notion that animals developed along evolutionary lines

B. Fossil record

1. Evidence from the fossil record is abundant and convincing
2. The fossil record documents the progressive changes that have occurred in previous geologic eras
3. The succession of fossil plants and animals is compatible with ideas about ancestry and relationships developed from other branches of science
   a. Evidence from biochemistry, molecular biology, and cell biology has established ***prokaryotic*** organisms as the ancestors of all life; prokaryotes, unicellular organisms that lack membrane-bound organelles, also are the oldest known fossils
   b. Similarly, the progressive emergence of vertebrates in the fossil record (first fish, then amphibians, reptiles, birds, and mammals) is the same sequence predicted from other bodies of evidence

C. Classification systems

1. Scientists and other observers (dating as far back as the ancient Greeks) have devised various systems of classifying organisms based on common characteristics
2. Animals commonly are grouped by shared anatomic and physiologic traits (see Chapter 4, Classification and Phylogeny of Animals)
3. Recently, genetic analysis has confirmed that animals grouped according to these criteria share a common ancestry

D. Geographic distribution

1. The geographic distribution of species is evidence for common descent
2. Oceanic islands contain many species of plants and animals that are found nowhere else in the world, yet these organisms are closely related to species on nearby continents
   a. The tropical animals of South America are more closely related to species of South American deserts than to species of the African tropics
   b. Australia contains a large native population of ***marsupial*** animals (animals whose young are born at an early stage of embryonic development; development is completed in maternal pouches), whereas the other continents are dominated by ***placental*** mammals (animals whose young are born at a more advanced developmental stage; embryonic development is completed in the uterus); this difference is attributed to Australia's geographic isolation (similar to an island)

E. Embryonic development

1. Closely related organisms go through similar stages of embryonic development

2. For example, all vertebrates have paired pharyngeal gill pouches on the sides of their throats at a similar stage of embryonic development
3. Vertebrates also display many other similarities during embryonic development

F. Molecular biology

1. Analysis with techniques of molecular biology shows that species closely related on the basis of other evidence have a greater similarity in their genes and proteins than do less closely related species
2. Neutral changes in the sequence of DNA bases (those that do not significantly change the activity of the gene product) are thought to accumulate at a relatively constant rate that is independent of natural selection; the number of neutral differences between equivalent gene sequences in two species is an approximate measure of the time since they had a common ancestor

IV. Microevolution

A. General information

1. Microevolution refers to changes in the gene frequencies (and therefore the frequencies of particular traits) in individual populations
2. The shift in gene frequencies often is due to the action of natural selection on phenotypic variation (the visible physical characteristics of an organism) within populations

B. Mechanisms of microevolution

1. All of the ***alleles*** (alternate forms of a gene that are expressed as various phenotypes) possessed by all the individuals in a population are collectively known as the ***gene pool***
2. Although mutations occur, the primary source of genetic variation in the next generation is the recombination of existing alleles during sexual reproduction
3. Natural selection acts on this population of variable individuals, favoring the survival of those with the most favorable combinations of traits; this action produces evolutionary change
4. Unless disturbed by outside factors, gene frequencies and phenotypic ratios in a population tend to remain constant; this is known as the *Hardy-Weinberg equilibrium*
   a. The Hardy-Weinberg equilibrium, proposed by G.H. Hardy and W. Weinberg in 1908, is expressed in a mathematical formula that can be used to calculate gene frequencies of a population over successive generations
   b. The formula is: $p^2 + 2pq + q^2$, where p represents the frequency of the dominant allele and q represents the frequency of the recessive allele
5. Five factors (agents of microevolution) can disturb the Hardy-Weinberg equilibrium: genetic drift, mutations, nonrandom mating, a changing population, and natural selection
   a. ***Genetic drift*** refers to changes in gene frequencies that occur as a result of chance (as opposed to those caused by selection, mutation, or migration); it often occurs when a small number of individuals become isolated from the main population
      (1) Because only a fraction of the population is represented, the potential gene pool for the next generation is greatly reduced

(2) This reduces genetic variability and changes the frequencies of alleles in the next generation, thereby disturbing the Hardy-Weinberg equilibrium

(3) Two causes of genetic drift—the bottleneck effect and the founder effect—have been observed

(a) In the *bottleneck effect,* a small population is isolated after an environmental disaster, such as a flood, earthquake, or fire; the population of survivors usually is not a representative sample of the original gene pool, and genetic drift ensues

(b) In the *founder effect,* a few individuals colonize an isolated habitat, such as an island; the smaller the founder population, the less representative it is of the original gene pool

(c) Because descendants of isolated populations (whether through the bottleneck or founder effect) lack the full genetic complement of the original population, they may be less adaptable and more vulnerable to changes in their environment

(d) Alternatively, when founder individuals are introduced to a new habitat, they may escape the pressure of predation or limited resources that formerly held the population in check; this can lead to phenomenal population explosions, as occurred with the spread of starlings and house sparrows across the United States

b. *Mutations* also can disturb the equilibrium; however, mutation rates usually are low and do not have much effect on gene frequencies from generation to generation

c. *Nonrandom mating,* in which individuals preferentially select mates with certain physical traits, disrupts the equilibrium; for the Hardy-Weinberg equilibrium to hold, random mating, with regard to the trait in question, must take place within the population

d. A *changing population* is one in which individuals enter or leave the population; the Hardy-Weinberg equilibrium describes gene frequencies in a stable population

e. Except for random change, *natural selection* is the only factor that can produce adaptive changes within populations

(1) Through natural selection, populations can respond to environmental change

(2) Natural selection acts on the phenotype, causing a shift in phenotypic ratios, thereby altering genotypic ratios (the frequency of dominant and recessive alleles in the population) in succeeding generations

V. Speciation

A. General information

1. A ***species*** is a group of organisms that interbreed freely to produce fertile offspring

2. Speciation often results when isolated populations within a species (such as those that result from the bottleneck or founder effect) do not have the opportunity for random mating with the parent population

a. Populations can be isolated by glaciation, seismic activity, desertification, or changing river beds

b. Differences between the populations accumulate as a result of chance and natural selection; genotypic ratios change quickly in the isolated population, leading to speciation

c. Speciation caused by geographic isolation is called *allopatric speciation,* which is the most common mechanism of speciation

d. Through allopatric speciation, single species often give rise to multiple new species in a process known as *adaptive radiation;* adaptive radiation populated the continent of Australia with marsupial animals and produced 14 species of finches on the Galapagos islands

B. Reproductive isolation

1. In nature, individuals of two different species do not typically interbreed; this is known as *reproductive isolation*
2. Physiologic, behavioral, or environmental barriers can prevent two species from interbreeding
3. Barriers to interspecies reproduction are called *isolating mechanisms;* they are categorized as prezygotic or postzygotic, depending on their timing in the reproductive process

   a. *Prezygotic isolating mechanisms* prevent mating between individuals of different species or prevent successful fertilization if mating does take place

   (1) Two species that live in different habitats within the same geographic area usually do not interact (called *ecologic isolation*)

   (2) Species with different activity cycles or breeding seasons will not interbreed (called *temporal isolation*)

   (3) Most species use complex species-specific chemical and behavioral signals during the mating process; mismatched signals prevent interbreeding (called *behavioral isolation*)

   (4) Different species, even within the same genus, may be sufficiently anatomically different to prevent interbreeding (for example, they may have different copulatory appendages) (called *mechanical isolation*)

   (5) The gametes of most species are immunologically incompatible; identifying proteins on the cell membranes of a sperm or egg may adhere only to corresponding molecules on the cell membrane of the other gamete (called *gametic isolation*)

   b. *Postzygotic isolating mechanisms* occur after fertilization and prevent the completion of embryonic development or reduce the reproductive capability of the offspring

   (1) *Hybrid sterility* describes the offspring of two species that are viable but sterile; a familiar example is the mule (the offspring of a horse and a donkey)

   (2) *Hybrid mortality* is high in interspecies embryos; hybrid ***zygotes*** usually do not complete embryonic development

   (3) *Reduced viability* is common in hybrid individuals; they may be poorly adapted to both parental habitats (for example, they may exhibit decreased predation efficiency or altered camouflage coloration)

VI. Macroevolution

A. General information

1. ***Macroevolution*** refers to evolutionary change above the species level (such as the genus, family, or order levels)
2. Macroevolution researchers are interested in major evolutionary events, including the origin of unique anatomic structures (such as bird feathers), trends in the fossil record, adaptive radiations, mass extinctions, and the rate of evolutionary change in different taxonomic groups

B. Mechanisms of macroevolution

1. *Preadaptations* (variations on existing anatomic structures) may give rise to unique anatomic characteristics
   a. A preadapted structure is one that evolved to serve one function but is converted to perform another
   b. Examples include bird feathers (which evolved from reptile scales) and the mammalian middle ear (which is a modified jaw element of reptiles)
2. In most animal embryos, different body parts grow and develop at different rates (called ***allometric growth***)
   a. Even slight changes in developmental rates can produce large anatomic differences
   b. For example, as humans develop, arms and legs grow faster than the head and trunk; in a newborn infant, the head comprises 25% of the body length but only half that in an adult
3. Genetic changes that alter the timing of development also produce novel features; this is known as ***paedomorphosis***
   a. Juvenile characteristics may be preserved in the adult, creating a novel organism
   b. For example, many salamanders retain their larval gills after they reach adulthood

---

Study Activities

1. Create a chart that compares and contrasts Darwin's theory of natural selection with previous ideas on evolutionary change.
2. Using the basic principles underlying Darwin's theory of natural selection, explain how strains of bacteria become resistant to specific antibiotics.
3. Explain and give examples of four different categories of evidence for evolution.
4. Compare the mechanisms of microevolution with those of macroevolution, and explain how each affects speciation.
5. Using your understanding of the principles of allopatric speciation, natural selection, and macroevolution, suggest various mechanisms through which species might arise.

3

Body Structure and Function

Objectives

After studying this chapter, the reader should be able to:
- Recognize the levels of organization of living organisms.
- Distinguish among the types of animal symmetry.
- Distinguish among the types of body cavities found in animals.
- Compare the types of skeletal support found in animals.
- Describe methods of homeostatic regulation (including osmoregulation, internal transport systems, gas exchange, nervous functions, and temperature regulation).

I. Levels of Organization

A. General information

1. To survive, all living organisms must perform certain basic functions, such as obtaining nutrition, reproducing, and excreting wastes
2. Living organisms are organized into levels of increasing complexity according to the way in which they accomplish these tasks

B. Levels of complexity

1. Five levels of complexity (organization) generally are recognized
2. The *protoplasmic level of organization* is present in unicellular organisms; all body functions are performed within the single cell
3. The *cellular level of organization* refers to groups of cells with some division of labor (for example, some cells may be responsible for nutrition and others for reproduction); colonial protozoans and sponges (phylum Porifera) are at the cellular level of organization
4. At the *tissue level of organization,* the cell groupings (called ***tissues***) are organized into layers of cells that have similar structures and functions
   a. The tissue layers are held together by sticky cell coatings or woven together with extracellular fibers
   b. Members of phylum Cnidaria and Ctenophora are at the tissue level of organization
   c. Animals at the tissue level of organization and beyond collectively are called ***eumetazoa,*** a category that comprises all of the animal phyla except Porifera
   d. There are four types of eumetazoan tissues

(1) *Epithelial tissue* lines the internal and external body surfaces and functions as a barrier against injury, invading microorganisms, and fluid loss

(2) *Connective tissue* binds and supports the other tissues and is characterized by small cell populations scattered throughout a nonliving extracellular matrix; examples include adipose tissue, blood, cartilage, and bone

(3) *Muscle tissue* is composed of contractile cells and functions in movement

(4) *Nervous tissue* senses stimuli and transmits signals from one part of the body to another

5. At the *organ level of organization*, the different tissues are grouped into specialized centers of function called ***organs***
    a. Organs usually are composed of more than one tissue type
    b. The first phylum to exhibit this level of organization is Platyhelminthes (flatworms), which have organs for vision, digestion, and reproduction
6. At the *organ system level of organization,* groups of organs work together to perform a basic metabolic function; the circulatory, digestive, and respiratory systems are organ systems

C. Embryonic tissue layers

1. The embryonic tissue layers of eumetazoans are called ***germ layers***
2. All organs and body structures develop from these embryonic layers
3. Three primary germ layers are found in most eumetazoans; animals with three germ layers are known as ***triploblastic***
4. The germ layers originate from folding and invagination during early embryonic development
    a. The innermost layer is called ***endoderm;*** it gives rise to the epithelial tissue lining the digestive tract and digestive organs
    b. The central layer is called ***mesoderm;*** it gives rise to the skeleton, muscles, circulatory system, and many other organs
    c. The outermost is called ***ectoderm;*** it gives rise to the epidermis and nervous systems
    d. The radiata (phylum Cnidaria and Ctenophora) lack mesoderm; they are referred to as ***diploblastic*** (having only two primary germ layers)

II. Animal Body Plans

A. General information

1. An animal's body plan is based on its general form, or geometry
2. The arrangement of body structures in relation to some axis of the body is called *symmetry*
    a. Animals with ***spherical symmetry*** have a body shaped like a ball; similar body parts are arranged around a central axis, and an infinite number of planes through this axis yields mirror-image halves
    b. In ***radial symmetry,*** similar body parts are arranged around a central axis, and more than one plane (but not an infinite number) through this axis yields mirror-image halves

c. ***Biradial symmetry*** is a modified form of radial symmetry; one body part is single or paired and thus only two planes passing through the central axis yield mirror-image halves

d. Animals with ***bilateral symmetry*** have a body plan in which the right and left sides are mirror-images

3. Spherical and radial symmetry commonly are found in drifting or *sessile* animals (those that remain fixed to one spot and usually do not move around)
4. Bilateral symmetry usually is found in animals with controlled mobility; such animals typically exhibit ***cephalization,*** a concentration of nerve cells and sense organs in the head area

B. Body cavities

1. Animals with bilateral symmetry are grouped according to the presence or absence of a body cavity (called a *coelom*)
2. The body cavity is a large, fluid-filled space between the body wall and the digestive tract
3. Animals with a body cavity have a tube within a tube design (the outer tube being the body wall and the inner tube being the digestive tract)
4. The evolution of a body cavity was a major advance in animal architecture
   a. The coelom provides much greater body flexibility than is possible for animals with a solid body structure
   b. The fluid-filled cavity serves as a mechanical buffer or cushion between the two body tubes
   c. The additional space allows for the formation and augmentation of new body structures
   d. The fluid is a medium for circulation and also provides skeletal support
5. ***Acoelomate*** animals do not have a body cavity surrounding the gut
   a. The area between the body wall and the digestive tract is filled with a mass of spongy cells (called *parenchyma*) that are derived from mesoderm
   b. Members of phylum Platyhelminthes (flatworms) and Nemertea (ribbon worms) are acoelomate animals
6. Animals that have body cavities are classified according to the embryonic origin of the tissues that give rise to the coelom
   a. ***Pseudocoelomate*** animals have a body cavity derived from the ***blastocoel*** (a fluid-filled cavity within the *blastula*, an early stage of embryonic development)
   b. ***Eucoelomate*** animals possess a *true coelom,* a body cavity derived from and lined by mesoderm
      (1) The true coelom is completely enclosed by a mesodermal lining called the *peritoneum*
      (2) The body organs are suspended from and surrounded by the *mesenteries* (thin sheets of connective tissue attached to the peritoneum)

C. Segmentation

1. Many animals have bodies that are arranged in a repetitive linear sequence of similar parts
2. This arrangement is known as ***segmentation*** or ***metamerism,*** and each segment is called a ***metamere*** or ***somite***
3. Segmentation can include both internal and external body structures, such as muscles, blood vessels, nerves, and locomotor structures

4. Segmentation is found in only three phyla: Annelida (segmented worms), Arthropoda, and Chordata

D. Body covering

1. The outer body covering of animals is known as the ***integument;*** it includes the skin and associated structures (such as hair and glands)
2. The integument functions as a barrier against injury, invading microorganisms, ultraviolet radiation, and fluid loss
3. In addition to its mechanical functions, the integument has several physiologic functions
   a. It plays a role in temperature regulation
   b. In some species, the integument functions in respiration and excretion
   c. The skin contains sensory receptors that provide essential information about the environment
   d. Skin secretions provide olfactory cues for intraspecies behavior (for example, sexual attraction or territorial marking)

E. Skeletal support

1. Almost all animals have some form of skeletal support
2. Skeletons maintain body shape, support body tissues, provide surface area for muscle attachment, and protect body organs
3. Not all skeletons are rigid; many invertebrates use their coelomic fluids for internal support
   a. These ***hydrostatic skeletons*** work because liquids are incompressible and can adjust to any configuration
   b. Because of these two properties, coelomic fluids transmit pressure changes rapidly and equally around the body
   c. Movement is accomplished by muscle contractions, which exert force (pressure waves) against the coelomic fluids
      (1) Layers of circular and longitudinal muscles in the body wall contract alternately
      (2) Contraction of muscles at the posterior (hind) end forces body fluids forward, extending the front of the animal's body
      (3) Muscle contractions then move the posterior end forward
      (4) The process is repeated in sequence to move the animal along
      (5) This method of locomotion is used by many worms
   d. A similar hydrostatic system is used to protrude specific body parts, such as the tube feet of echinoderms or the siphons of clams
4. Rigid skeletons can be either internal (*endoskeletons*) or external (*exoskeletons*)
5. *Vertebrates* have an endoskeleton with a vertebral column (backbone); *invertebrates,* even those with endoskeletons, do not have backbones
6. There are several structural and physiologic differences between endoskeletons and exoskeletons
   a. Exoskeletons are composed of nonliving organic and inorganic molecules
      (1) Exoskeletons may be composed of the nitrogenous polysaccharide ***chitin,*** protein, calcium carbonate, or a combination of these substances
      (2) Exoskeletons are secreted by cells of the integument; in other words, they are of ectodermal origin
      (3) The exoskeleton may be rigid (as in snails) or articulated (jointed and movable) as in arthropods and bivalve molluscs

(4) Because the exoskeleton is nonliving, it does not grow with the organism; it must be shed (***molted***) and replaced to permit increased body size
(5) Exoskeletons are common in the phyla Mollusca and Arthropoda

b. Endoskeletons are formed inside the body and usually are of mesodermal origin
(1) As with exoskeletons, they are composed of organic and inorganic molecules
(2) Endoskeletons range from networks of mineral spicules found in flatworms (phylum Platyhelminthes) to the interconnected calcareous plates of echinoderms and the articulated cartilage or bony skeletons of vertebrates
(3) Endoskeletons are composed of living cells and thus grow with the organism
(4) In addition to its support function, vertebrate bone also stores and releases calcium and phosphorus; in higher vertebrates, such as mammals, blood cells are produced in the bone marrow

c. Chordate skeletons may be composed of cartilage or bone; nonvertebrate chordates have only a supportive longitudinal rod called a ***notochord***
(1) Cartilage is the principal skeletal element of some vertebrates
(a) The jawless fishes (class Agnatha) and the sharks and rays (class Chondrichthyes) have skeletons composed solely of cartilage
(b) The cartilage cells, called *chondrocytes*, secrete a tough but flexible protein matrix
(2) Animals from other vertebrate classes have skeletons composed primarily of bone; skeletal elements of cartilage are found in locations where flexibility or smooth articulation is desirable
(a) The skeletons of vertebrate embryos are composed of cartilage, which is progressively replaced by bone in most locations during development
(b) Bone cells, called *osteocytes*, secrete an extracellular matrix laid down in concentric layers
(c) The matrix is composed of protein fibers and inorganic calcium salts

F. Locomotion

1. There are four basic methods of animal locomotion
a. ***Amoeboid movement,*** found in single cells, is accomplished by the extrusion of protoplasmic extensions called *pseudopodia*
b. Single cells also may move by the beating of ***cilia*** and ***flagella***
c. Hydrostatic propulsion is used by many animals without rigid skeletal support (such as worms)
d. Other animals move by means of specialized appendages, such as tentacles or legs
2. ***Aquatic*** animals must overcome the viscosity and dynamic qualities of water in order to move; animals that move rapidly through the water have streamlined bodies to reduce drag
3. Some animals perform voluntary movements by means of bands of skeletal (striated) muscles

a. These muscles contract as protein filaments called actin and myosin slide past each other
b. Muscle fibers are organized into compact bundles or bands that are anchored to the skeleton

III. Homeostatic Functions

A. General information

1. ***Homeostasis*** (maintaining a steady state) is accomplished by the coordinated activities of all body systems
2. A fundamental function of animal homeostasis is control of the internal fluid environment
   a. All body cells are surrounded by protective extracellular fluids; the solute composition of these fluids, known as the *osmotic concentration,* remains relatively constant
   b. Production of metabolic wastes and necessary exchanges with the external environment tend to disrupt the delicate balance of water and solutes in body fluids
   c. The metabolic processes that preserve this solute and fluid balance collectively are known as ***osmoregulation***

B. Osmoregulation

1. Osmoregulation is directly related to environmental conditions and thus varies among ***marine*** (ocean-dwelling) animals, freshwater animals, and terrestrial animals
2. The composition of the body fluids of most invertebrates is similar in overall osmotic concentration and ionic composition to that of sea water
   a. Most marine invertebrates are in *osmotic equilibrium* with their seawater habitat; their body fluids are approximately ***isotonic*** to sea water
   b. Although no animal's body fluids are exactly isotonic with sea water, marine invertebrates do not encounter the severe osmoregulatory problems experienced by freshwater and terrestrial animals
   c. The bodies of marine invertebrates are permeable to salt and water; the osmotic concentration of their body fluids varies with changes in environmental salinity
   d. Because marine invertebrates are unable to regulate the osmotic concentration of their body fluids (except for minute, fine-tuning changes), they are referred to as ***osmoconformers***
   e. Most osmoconformers have a limited ability to osmoregulate and therefore are ***stenohaline,*** or restricted to a narrow range of salinities
   f. Animals that can survive a wider range of salinity changes are called ***euryhaline;*** typical examples are arthropods (such as crabs) that inhabit saltwater marshes, estuaries (where rivers and oceans meet), and coastal intertidal zones (which alternately are immersed and exposed by rising and falling tides)
      (1) These habitats have highly variable salinity levels
      (2) The body fluids of euryhaline invertebrates are ***hypertonic*** to the dilute seawater of the coastal environment; their osmotic problems include influx of water and loss of body salts

(3) Osmotic flexibility allows these animals to inhabit locations that are unstable but extremely rich in resources
(4) In seawater, euryhaline invertebrates are osmoconformers; as the seawater becomes progressively more dilute, they function as ***osmoregulators*** (they have a limited ability to maintain the osmotic concentration of their internal body fluids)
(5) Although the osmotic concentration of their body fluids changes in response to varying environmental salinity, the change occurs slowly and to a lesser degree than the external changes
(6) Excess water is excreted via the kidneys; ions are actively removed from the seawater and transported, against the concentration gradient, into the blood

3. All freshwater and terrestrial animals are osmoregulators; they maintain their internal body fluid concentrations regardless of external conditions
4. The bodies of freshwater animals (vertebrates and invertebrates) are strongly hypertonic to their environment; their osmotic problems include influx of water and loss of body salts
   a. The body covering is relatively impermeable; for example, fish are covered with scales and mucus and arthropods have tough, chitinous exoskeletons
   b. The kidneys excrete a very dilute urine to remove excess water
   c. Replacement ions are obtained in food and by special salt-absorbing cells on the gills
   d. The ionic concentration of internal body fluids is maintained at the lowest level possible for the animal's metabolism; this reduces the energy required to maintain an osmotic balance
5. Modern marine bony fishes (class Osteichthyes) are descendants of freshwater bony fish that returned to the sea during the Triassic period, approximately 200 million years ago
   a. Consequently, their body fluids are ***hypotonic*** to sea water; osmotic problems include dehydration (similar to terrestrial animals) and influx of salts
   b. These problems are exactly the opposite of those experienced by freshwater animals
   c. To compensate for water lost through diffusion, marine fishes drink seawater
   d. Excess salts are removed via salt-secreting cells on the gills or excreted with the feces and urine
6. Terrestrial animals are exposed to air and thus must prevent evaporative water loss across the skin and respiratory surfaces
   a. The body covering is relatively impermeable
   b. Many terrestrial animals inhabit moist or humid habitats (such as burrows); many also are ***nocturnal*** (active at night)
   c. Respiratory water loss is reduced by internal gas exchange organs, such as the lungs
   d. Excretory water loss is reduced by internal excretory organs, such as the kidneys
      (1) The kidneys reclaim water and concentrate urine, reducing urine output
      (2) In some species, nitrogenous wastes are excreted in an insoluble form (which requires less water for elimination) such as ***uric acid*** (in snails, birds, insects, and reptiles)

C. Internal transport systems

1. As animals increase in size and metabolic rate, an efficient transport (circulatory) system is necessary to distribute essential materials and remove wastes
2. The type of circulatory system is related to the size, complexity, and life-style of the organism
    a. Some animals use water from the environment as a circulatory fluid
        (1) In sponges, the beating of flagellated cells circulates water through a series of channels and canals
        (2) In cnidarians, flagellated gastrodermal cells circulate water through the gastrovascular cavity
        (3) In echinoderms, the beating of cilia draws a current of water through the coelom
    b. Pseudocoelomate animals, such as nematodes and rotifers, circulate body cavity fluids
        (1) Most pseudocoelomates are very small
        (2) Contractions of the body wall move the pseudocoelomic fluids, which are in direct contact with internal tissues and organs
        (3) Some types of cells are present in the body fluids; their purpose is unclear, but they may function in transport and waste collection
    c. Most other animals have specialized circulatory systems, which contain a circulatory fluid (*blood*)
        (1) Blood is a viscous fluid; a significant amount of propulsive force is needed to overcome friction and to keep the blood moving through the system
        (2) The force is provided by the heart, contractile vessels, and skeletal muscles
    d. Circulatory systems have two basic designs, or structural plans
        (1) In ***open circulatory systems,*** blood is pumped through the heart, through *arteries*, into a series of chambers known as *blood sinuses*
            (a) The blood sinuses open into the body cavity, and fluid is in direct contact with the organs; gas exchange occurs by *diffusion* (passive exchange across the cell membrane)
            (b) The body cavity in open circulatory systems is known as the ***hemocoel;*** the blood (which also is the coelomic fluid) is called ***hemolymph***
            (c) Arthropods and most molluscs (except cephalopods) have open circulatory systems
            (d) Hemocoels also may function as hydrostatic skeletons and in temperature regulation
            (e) Blood pressure and the speed of circulation are lower in open circulatory systems compared with closed systems; open systems, therefore, are characteristic of less active animals
        (2) In ***closed circulatory systems***, blood is confined to vessels or vessel-like channels as it circulates through the body
            (a) Blood moves from the heart through *arteries* to broad networks of *capillaries* in the tissues and returns through the *veins*
            (b) Exchange of circulated materials takes place in the capillary beds; small molecules diffuse between the ***interstitial fluids*** (coelomic fluids surrounding the tissues and organs) and the blood in the capillaries

(c) Closed circulatory systems are characteristic of active animals, such as cephalopod molluscs and vertebrates

e. Open and closed circulatory systems have various types of pumping organs

(1) These include *contractile blood vessels* (found in annelids), *tubular hearts* (found in most arthropods), and *chambered hearts* (found in molluscs and vertebrates)

(2) Contraction in muscles of the pumping organ is initiated by two methods

(a) In ***myogenic*** contraction, the heartbeat originates within the cells of the pumping organ; this system is found in molluscs and vertebrates; if removed from the body, a myogenic heart continues to beat

(b) In ***neurogenic*** contraction, the heartbeat is initiated by nerve action; this system is found in arthropods and in the contractile blood vessels of annelids

D. Gas exchange

1. A primary function of circulatory systems is gas exchange; oxygen is transported to body cells and carbon dioxide (a waste product of cellular metabolism) is removed
2. Oxygen and carbon dioxide are transferred exclusively by diffusion
   a. Gases must be in solution to diffuse across a cell membrane
   b. Diffusion depends on the concentration gradient of the gases at the exchange site; the gradients are maintained by constant circulation of internal fluids
3. Many small aquatic invertebrates exchange gases across the skin and have no special respiratory structures
   a. This type of gas exchange is feasible only in animals (such as protozoans, sponges, cnidarians, and many worms) that have very high surface-area-to-volume ratios
   b. Multicellular animals with this type of respiration increase their surface area (relative to body mass) with thin, flattened profiles or tubular shapes that provide maximal exposure for gas transfer (on both the inner and outer body surfaces)
4. Most aquatic animals have gills, which are specialized for gas exchange
   a. Gills are highly vascularized organs composed of many thin plates or filaments that extend from the body surface and provide increased surface area for diffusion
   b. Water is moved over the gills by the beating of cilia or other appendages, by forward locomotion of the animal, or by muscular pumping action
   c. In molluscs and fishes, diffusion is facilitated by a *counter-current exchange system*, in which the direction of blood flow within the gills is opposite that of water flow over the gills (see Chapter 16, Vertebrate Physiology)
5. Terrestrial vertebrates have internal lungs for gas exchange
   a. Lungs are highly vascularized internal cavities that are kept moist by body fluids
   b. Air is moved into and out of the lungs by muscular action
   c. Ventilation of internal lungs is difficult to accomplish

      (1) In contrast to the unidirectional flow of water across fish gills, air must enter and exit a lung through the same opening

(2) The air in lung passages is never completely expelled; a reservoir of dead air remains with each breath; normally only 1/6 of the air in the lungs is replaced with each breath

(3) The physical nature of air helps to resolve the problems of gas exchange in terrestrial vertebrates

(a) The oxygen content of air is much higher than that of water; air-breathing animals can move a smaller volume of air over the lungs and still obtain sufficient oxygen for their respiratory needs

(b) Air is much less dense than water and therefore can be moved with less energy expenditure

6. Birds have a modified, highly efficient respiratory system
   a. It consists of anterior and posterior air sacs interconnected with tubes and provides unidirectional air flow through the lungs
   b. This system provides birds with the large quantities of oxygen needed for flight
7. In insects, gases diffuse through tracheal tubes directly to the body tissues; blood plays little or no role in gas exchange
8. To increase the blood's capacity to transport oxygen, many animals have carrier molecules called ***respiratory pigments***
   a. Pigment molecules, which are protein molecules, may be dissolved in blood or body fluids or contained within specific blood cells
   b. In the presence of high oxygen concentrations, pigment molecules combine with oxygen

(1) At the site of gas exchange, oxygen concentrations are high relative to body fluids, and the pigment molecules bind to oxygen

(2) When the oxygen concentration is low relative to body fluids, as in the tissues, the pigments release oxygen

(3) In addition to carrying oxygen from loading to unloading sites, some pigments carry oxygen reserves that are released only when tissue oxygen levels are very low

(4) Factors such as temperature and carbon dioxide concentrations affect the oxygen-carrying capacity of respiratory pigments

   c. There are several types of respiratory pigments

(1) ***Hemoglobin*** is a large iron-containing molecule that is found in almost all vertebrates and many invertebrates

(2) ***Hemocyanin*** contains copper (rather than iron) and always is in solution (never within blood cells); it is found in many molluscs and arthropods (especially crustaceans)

(3) ***Hemerythrin*** and ***chlorocruorin,*** which both contain iron, are less common respiratory pigments; they are found only in a few groups of invertebrates (mostly worms)

E. Nervous and sensory functions

1. With the exception of sponges and some parasites, all animals have a nervous system that receives sensory impulses and responds to external stimuli
2. The functional units of a nervous system are the nerve cells, or ***neurons***
3. In general, the more complex the animal, the more complex the nervous system; even simple nervous systems, however, provide animals with a broad spectrum of behavioral options

a. The simplest system is the cnidarian ***nerve net,*** a loose organization of interconnected neurons
   (1) Transmission of nerve impulses is *nonpolar*, meaning that the nerve impulses are propagated in all directions around the body simultaneously
   (2) There are no specialized sensory, motor, or connector neurons
b. In bilaterally symmetrical animals, aggregations of nerve cells at the anterior end of the body form a brain
   (1) The brain (or ***cerebral ganglia***) integrates sensory information and motor responses for the body
   (2) Longitudinal strands of nerve cells, called *nerve cords*, extend from the brain to the posterior end of the body
   (3) Most invertebrates have two ventral nerve cords, while vertebrates have one dorsal nerve cord
   (4) Distinct *afferent* (sensory) and *efferent* (motor) neurons, with connectors called *interneurons,* integrate sensory signals and coordinate motor responses

4. Animals have several different types of sensory receptors
   a. ***Tactile receptors*** are sensitive to touch and vibration and include a variety of hairs, bristles, and spines; they usually project outward from the body surface
   b. ***Georeceptors*** (organs of equilibrium and balance) are sensitive to the pull of gravity and to low-frequency vibrations
      (1) In invertebrates, the most common georeceptors are called ***statocysts***
         (a) Statocysts consist of a fluid-filled chamber; the epithelium of the chamber is covered with touch-sensitive hairs
         (b) A solid granule or pellet called a *statolith* is suspended in the fluid
         (c) When the animal changes position, the statolith shifts position and touches the sensory hairs
      (2) In vertebrates, equilibrium is maintained by the *labyrinth*, which is located in the inner ear
         (a) The labyrinth consists of two small chambers (the *sacculus* and *utriculus*) and two or three *semicircular canals*
         (b) The sacculus and utriculus are similar in structure and function to statocysts
         (c) The semicircular canals respond to body rotation
   c. *Auditory receptors* are found in many animals
      (1) Among the invertebrates, insects have the most complex auditory receptors
      (2) Vertebrate auditory receptors, or ears, also are well developed; sound plays an important role in the lives of most vertebrate animals
   d. ***Chemoreceptors*** are sensitive to chemicals; they are universal among animals, although the range and quality of reception varies widely
      (1) The types of chemicals to which particular animals respond are closely associated with their life-style
      (2) Insects and other animals produce species-specific chemicals called ***pheromones***, which play a role in a complex chemical language
      (3) *Taste buds* are important in identifying and selecting acceptable food items
   e. ***Photoreceptors*** are sensitive to light and are found in most animals

(1) All photoreceptors have light-sensitive pigments that absorb light energy in the form of photons; this absorbed energy stimulates the sensory neurons of the photoreceptor

(2) Photoreceptors range from simple to complex; protozoans have organelles called *stigmata,* which are simple spots of light-sensitive pigment

(3) Photoreceptors in multicellular animals are grouped into three general types

(a) *Ocelli* (sometimes called *simple eyes* or *eyespots*) are the least complex and provide information on light intensity that is important to orientation, protective changes in color, and timing of reproductive cycles; ocelli are common in several phyla, including cnidarians and arthropods

(b) ***Compound eyes***, common in arthropods, are a collection of separate units called *ommatidia;* each has its own field of vision and its own nerve

(c) ***Complex*** or ***camera eyes*** are typical of vertebrates and cephalopod molluscs; light impacts photoreceptors called *rods* and *cones* in the retina of the eye

F. Temperature regulation

1. Temperature control is an important part of homeostatic regulation
   a. Metabolic reactions are very temperature sensitive; the rate of most enzyme-catalyzed reactions doubles with each 10°C rise in body temperature
   b. If body temperature drops too low, metabolic processes are inhibited
2. Animals have two basic methods of stabilizing their body temperature: behavioral thermoregulation and metabolic heat
   a. ***Ectotherms,*** or ***poikilotherms,*** use behavioral thermoregulation to control their body temperature
      (1) Ectotherms maintain a constant body temperature by moving into and out of the sun
      (2) Ectotherms also can adjust their rate of metabolic reactions in response to environmental temperature
      (3) Ectothermic animals include invertebrates, most fish, amphibians, and reptiles
   b. ***Endotherms,*** or ***homeotherms,*** use metabolic heat to maintain a stable body temperature
      (1) A large percentage of an endotherm's daily caloric intake is used to generate body heat, especially in cold weather
      (2) For this reason, endotherms eat more than similarly sized ectotherms
      (3) Endothermic animals include birds, mammals, and some fish
3. Desert animals show behavioral and physiologic adaptations for high temperatures
   a. Many are nocturnal or fossorial (live in burrows underground)
   b. Large desert animals have a light body color (to reflect light) and excellent insulation (body hair or concentrations of fatty tissue on the back)
   c. Sweating is minimal to reduce water loss
   d. Some desert animals, such as the kangaroo rat, can obtain all the water they need from dry vegetation; they may never drink water their entire lives
4. Birds and mammals have several strategies to keep warm in cold climates

a. They have effective insulation, such as fur, feathers, or the thick layer of fatty tissue (blubber) found in marine mammals
b. Warm blood is shunted away from the extremities to conserve body heat
c. Some animals may spend the winter in tunnels and chambers excavated under the snow; through this semi-fossorial existence, they avoid exposure to the severe cold above the insulating snow
d. In extreme cold weather, small birds and mammals may undergo daily ***torpor*** (a severe drop in body temperature when inactive or asleep)
e. Larger mammals may enter ***hibernation*** in winter; in this prolonged dormant period, the animals live on stored fat reserves

Study Activities

1. Name and describe the five levels of organization of living organisms.
2. Distinguish among the types of animal symmetry.
3. Describe the types of body cavities found in animals.
4. Compare the types of skeletal support found in animals.
5. Describe the methods of homeostatic regulation in animals (including osmoregulation, internal transport systems, gas exchange, nervous functions, and temperature regulation).

4

Classification and Phylogeny of Animals

Objectives

After studying this chapter, the reader should be able to:
• Compare the major goals of taxonomy and systematics.
• Describe the Linnaean system of classification.
• Become familiar with the levels of the taxonomic hierarchy.
• Distinguish among the three principal methods of animal classification.
• Compare homologous and analogous characters and explain how they are used in animal classification.
• Describe the five kingdom system of classification.

I. History of Classification

A. General information

1. A classification system is necessary to efficiently inventory the huge number of species of organisms on earth
   a. To date, about 2 million species have been named and described
   b. Biologists estimate that 4 to 30 million unidentified species remain
   c. In order to logically study this diverse array of organisms and the relationships among them, some sort of system for naming and arranging them is necessary
2. ***Systematics*** is the study of the diversity of living organisms and the relationships among them; systematics includes ***taxonomy,*** which is the identification and classification of species

B. Development of a classification system

1 The modern scheme of classification was developed and refined by Carolus Linnaeus, a Swedish botanist, in the 1700s
   a. Linnaeus was so effective in setting up a functional system for classifying plants and animals that he often is called the father of taxonomy
   b. Modern taxonomy started with his two books, *Species Plantarum* (on plants) in 1753 and *Systema Naturae* (on animals) in 1758
2. Linnaeus developed a hierarchical system of classification
   a. Species are arranged into a series of increasingly comprehensive groupings
   b. This hierarchical system is still in use, although scientists now use more categories within the hierarchy

3. The taxonomic hierarchy currently in use has a descending series of seven main ranks, with subdivisions within each rank
   a. The seven ranks are (in descending order): kingdom, phylum, class, order, family, genus, and species
   b. Subdivisions include categories such as subphylum, superorder, and subspecies; at present, more than 30 taxonomic groupings are used to classify living organisms
   c. These additional divisions are necessary to express the complex nature of evolutionary relationships within the various groupings
4. Using the system developed by Linnaeus, each species is given a unique, two-part name; this system for naming species is called ***binomial nomenclature***
   a. This two-part Latinized name is the organism's scientific name
   b. Scientific names are standard worldwide and are the same in every language
   c. Scientific names are composed of two levels in the taxonomic hierarchy
      (1) The first part is the genus name, which always is capitalized
      (2) The second part, the specific name, always begins with a lowercase letter; because the scientific name is written in a foreign language (Latin), it always is italicized or underlined
      (3) For example, the scientific (or species) name for humans is *Homo sapiens*
5. No two organisms share the same scientific name
6. Species are defined biologically as a group of organisms that interbreed to produce fertile offspring; in nature, individuals of two different species typically do not interbreed

C. Basis of the classification system

1. Organisms in a phylum or other taxonomic grouping share similar characteristics because they share a common ancestry
2. Taxonomists try to express this evidence of relationship when classifying organisms
3. ***Characters*** are observable traits or features that biologists use to separate organisms and assign them to different taxonomic groupings
   a. Characters, the traits of an individual's ***phenotype*** (observable set of physical characteristics), are the visible evidence of an individual's ***genotype*** (total genetic or hereditary makeup of an organism)
   b. Sets of characters are used to define species and higher taxonomic groupings
4. Characters that are used to classify organisms must be homologous
   a. ***Homologous*** characters are features that are inherited from a common ancestor and show a similar pattern of embryonic development
      (1) Homologous characters show a degree of relationship in the groups that display them, but they may serve a similar or different function within the various groups
      (2) The homologous bones of the human forelimb, porpoise flipper, bat wing, and horse leg are alike in structure and embryonic origin but have different functions
   b. ***Analogous*** characters, those that are similar in function but not in origin, are not used to classify organisms; analogous characters arise through ***convergent evolution***, a process by which unrelated organisms inde-

pendently develop similar characteristics because they inhabit similar environments

II. Traditional and Modern Approaches to Classification

A. General information

1. Three different approaches to classification currently are used
    a. The traditional approach is based on evolutionary taxonomy
    b. The modern approaches are based on numerical taxonomy and phylogenetic systematics
2. The three systems feature elements of taxonomy (naming and categorizing organisms) and systematics (clarifying the evolutionary history of organisms)
3. There are pros and cons to each of the classification methods
4. Most modern taxonomists and systematists use a combination of techniques from these three methods, customizing their approach to the group that they are studying

B. Evolutionary taxonomy

1. ***Evolutionary taxonomy*** classifies species into groupings based on the number of shared homologies and evolutionary history
2. The goal of this system is to reconstruct evolutionary history as closely as possible
3. The branching diagrams that display this type of classification are called ***phylogenetic trees***
4. Phylogenetic trees display the passage of time on the vertical axis and the degree of evolutionary divergence on the horizontal axis
    a. The base of the phylogenetic tree represents the earliest ancestral species
    b. As the branches move farther from the base of the tree, they represent progressively more recent divisions of the various evolutionary lines
    c. Most phylogenetic trees are *monophyletic,* meaning that all organisms on that tree have descended from a common ancestor
5. The ancestor-descendant relationships used to construct a phylogenetic tree are established from the fossil record and from structural and biochemical homologies
    a. Recognition of specific characters as homologies may be difficult and relies heavily on the expertise of specialists
    b. Some taxonomists view this classification process as subjective and arbitrary; to address these concerns about the accuracy of this method, two other methods of classification have been developed

C. Numerical taxonomy

1. ***Numerical taxonomy (phenetics)*** measures as many characters as possible, then groups organisms on the basis of similarity
2. This classification system does not consider evolutionary history; it is based solely on the percentage of similar characters
3. Groupings of similar organisms are called *operational taxonomic units*
4. The branching diagrams that display this type of classification are called ***phenograms***

a. A phenogram is constructed through computer analysis of a large number of characters; usually more than 100 characters are analyzed

b. Analogous and homologous characters are not analyzed separately; numerical taxonomists believe that homologies (similarities due to common ancestry) always outweigh analogies (similarities due to common environment), if a large enough set of characters is analyzed

5. Phenetics was popular in the 1960s but is no longer commonly used

a. Hundreds of phenetic computer programs were written, but the same data, analyzed with different programs, produced different phenograms

b. Computer programs grew so complex that many taxonomists did not understand the numerical calculations and, therefore, could not assess their degree of accuracy

c. Also, phenograms do not account for evolutionary history or clarify common ancestry

D. Phylogenetic systematics

1. ***Phylogenetic systematics (cladistics)*** groups organisms entirely by the degree of common descent
2. Developed in the 1960s, this classification scheme is widely used
3. The branching diagrams that display this type of classification are called ***cladograms***

a. Each branch of a cladogram represents the division of a parent species into two daughter species

b. The branches show the degree of relationship but not the historical events; therefore, cladograms do not have a time scale

4. The degree of relationship between species is based on the percentage of shared derived characters

a. Homologous characters are considered to be either *primitive* (retained from a remote common ancestor or showing ancestral traits) or *derived* (having developed more recently)

(1) Examples of primitive characters within the primates are body hair, milk glands, and a placenta (traits shared within the ancestral phylum Chordata)

(2) An example of a derived character (unique to the primates) is the opposable thumb

b. The designations *primitive* and *derived* depend on the circumstances in which they are used; in other words, any trait can be considered primitive or derived depending on the group under consideration and its level in the taxonomic hierarchy

(1) The set of selected characteristics is compared between the group being classified (usually a closely related group, such as a species within a genus, or the genera within a family) and another, distantly related group, called an *outgroup*

(2) Any character that the related group shares with the outgroup is considered to be primitive

(3) Characteristics shared by the related group, but not present in the outgroup, are considered to be derived

(4) Groups are organized on a cladogram based on the sequence in which unique derived characters appear in members of the related group

5. Cladists claim that this method is less subjective and less influenced by individual bias than other methods of classification
   a. The method is based on prediction, analysis, and testing
   b. The cladist must be very specific in describing the characters to be analyzed and in identifying the group of related organisms to be classified
6. As with the other classification methods, cladistics has drawbacks; construction of a cladogram is complex and usually performed by a computer, thus many alternative constructions must be examined and eliminated to reach the final arrangement

III. Divisions of the Animal Kingdom

A. General information

1. Living organisms are organized into five kingdoms: Monera, Protista, Fungi, Plantae, and Animalia
   a. Kingdom Monera includes the prokaryotes, such as bacteria
   b. Kingdom Protista includes the unicellular eukaryotes, the protozoans, and the eukaryotic algae
   c. Kingdom Fungi includes heterotrophic organisms, such as fungi, yeasts, and molds
   d. Kingdom Plantae includes the multicellular autotrophs (green plants that perform photosynthesis)
   e. Kingdom Animalia includes the animals
2. The protozoans have many animal-like characteristics but are separated into the Kingdom Protista

B. Larger divisions of the animal kingdom

1. Although the phylum is the largest taxonomic level in practical usage, zoologists often combine phyla that have similar anatomic features and patterns of embryologic development
2. In some cases, members of these convenient groupings are united by common ancestry as well as by similar traits
   a. Eumetazoa is a collective term that includes all animals with tissues and organs
   b. Radiata is used to describe animals that have radial symmetry
   c. Bilateria encompasses all animals that have bilateral symmetry
   d. ***Protostomes*** are animals that have ***spiral cleavage,*** in which the mitotic divisions of the zygote (fertilized egg) occur in a spiral pattern
      (1) The mouth arises from the ***blastopore*** (the external opening in the ***gastrula*** [invagination] stage of embryonic development, which may develop into the mouth or the anus)
      (2) Protostomes can be acoelomates, pseudocoelomates, or eucoelomates
   e. ***Deuterostomes*** are animals that have ***radial cleavage,*** in which the mitotic divisions of the zygote are distributed evenly, such that the zygote can be divided along any plane to form two mirror-image halves
      (1) The mouth arises anterior to the blastopore (that is, in front of the blastopore)
      (2) Deuterostome animals are all eucoelomates

Study Activities

1. Describe the Linnaean system of classification.
2. Name the seven levels of the taxonomic hierarchy and arrange them in descending order.
3. Compare homologous and analogous characters and give examples.
4. Create a chart that compares and contrasts evolutionary taxonomy, numerical taxonomy, and cladistics.
5. Name and explain five types of biochemical and structural homologies that are used for animal classification.
6. Explain the five kingdom system of classification.

5

Phylum Porifera

Objectives

After studying this chapter, the reader should be able to:
- Describe the basic characteristics of sponges.
- Explain current hypotheses concerning the evolutionary ancestry of sponges.
- Differentiate among the three sponge body plans.
- Discuss the functions of the specialized cell types in sponges.
- Describe how sponges perform their basic life functions.
- Identify and characterize the three classes of sponges.

I. Basic Characteristics

A. General information

1. Members of the phylum Porifera, commonly called sponges, are the most primitive of the multicellular animals
2. Most authorities place sponges at the cellular level of organization because their modes of nutrition, gas exchange, and reproduction are similar to those of protozoans
    a. Sponges lack tissues and organs
    b. They have no head or anterior end, mouth, gut cavity, nervous system, or sense organs
    c. A few types of specialized cells function in nutrition and reproduction, but the cells are scattered and not well organized compared with those of most other animal groups
    d. Most body cells are extremely versatile, retaining the ability to move and to change form and function
3. Sponges are sessile animals; they remain fixed to one spot and usually do not move from it
4. Pores line the sponge body, giving the phylum its name; Porifera means "pore bearer"
    a. Sponge nutrition requires a continuous water current, drawn through the pores in the body wall by the beating of flagella
    b. Sponges filter an enormous quantity of water; a large sponge may move up to 1,500 liters of water per day

B. Ecologic relationships

1. About 9,000 living species of sponges have been identified

a. All live in water, primarily in marine environments
b. Sponges are found in all seas and at all depths, even in deep ocean trenches
2. These animals are filter feeders, removing bacteria and small organic particles from the water
3. They attach to and grow on many other animals, including cnidarians, molluscs, and arthropods (see Chapters 6, Radiate Phyla: Cnidaria and Ctenophora; 9, Phylum Mollusca; and 11, Phylum Arthropoda)
4. Many other invertebrates and fish live in or on sponges as commensals (organisms that obtain food or other benefits from host organisms without affecting the host); some use the sponge only as living space and for protection, but others depend on the sponge's water current to provide a constant source of suspended food particles
5. Sponges have structural and chemical defenses
a. More than 75% of sponges produce chemical toxins
b. The needle-like skeleton of many sponges also helps to deter potential predators, leaving them with a mouthful of sharp splinters

C. Evolutionary relationships

1. Sponges are considered primitive because they lack organs and have only a small number of different types of cells; however, they are highly specialized in ways that have helped them adapt to a sessile life-style
a. Sessile animals such as sponges tend to be radially symmetrical or ***asymmetrical,*** which enables them to attach to a variety of substrates
b. Circulation of water through the sponge's body allows for filter feeding
c. Most sponges are ***hermaphrodites*** (both sexes present in the same individual), which maximizes reproductive flexibility for a sessile animal
2. The sponge body plan and pattern of embryonic development are unique in the animal kingdom; these characteristics show that sponges diverged early from the evolutionary line that led to other multicellular animals
3. Current hypotheses of sponge origin favor ancestry from individual or colonial flagellate protozoans
a. Sponges are similar to individual flagellated protozoans; both feed by phagocytosis and both have flagellated larvae
b. Sponges also are similar to colonial flagellated protozoans
(1) The choanocytes, flagellated collar cells of sponges, resemble the collared cells of colonial choanoflagellates
(2) Inversion of the body wall, which occurs during the embryonic development of sponges, also occurs in the colonial flagellate *Volvox*

II. Sponge Form and Function

A. General information

1. Sponges are radially symmetrical or asymmetrical in shape
2. Their body structure is an organized arrangement of canals and chambers through which water flows
a. Surface pores lead into a central body cavity known as the *atrium* or *spongocoel*
b. Water enters the sponge through many microscopic openings called *incurrent pores* or *ostia*, which are found all over the body surface

(1) These openings are connected by a series of canals, which are lined with flagellated cells called *choanocytes*
(2) Each choanocyte beats independently
(3) Together, the beating of the flagella creates a unidirectional flow of water through the body cavity

c. The water stream, pumped by beating flagella, exits the body cavity by one or more large excurrent vents (openings in the body wall) called *oscula*

d. The choanocytes not only maintain the water current but also filter out small food particles

3. The sponge skeleton prevents the choanocytes from collapsing
   a. In most sponges, the skeleton consists of small crystalline spikes called *spicules,* which are composed of calcium carbonate or silicon dioxide
   b. Specialized cells secrete the sponge spicules
      (1) Spicules are microscopic and distributed throughout the body; in some species, they may be fused into long strands
      (2) The size and shape of the spicules varies and may be used to identify different species
   c. In class Demospongiae, the spicules may be augmented or replaced by a network of fibers called *spongin*
      (1) Spongin fibers are composed of ***collagen,*** the major structural protein of invertebrates
      (2) The fibers are woven into a thick network that provides structural support while retaining flexibility; the common bath sponge has a spongin skeleton

B. Types of sponge structures

1. Sponges are separated into three body types according to the layout and complexity of their canal systems (see *Types of Sponge Structures*)
2. *Asconoid* sponges have a simple vase-like structure
   a. The body is a radially symmetrical tube, with a thin body wall and a single osculum
   b. The ostia (incurrent pores) lead into the spongocoel (atrium), which is lined with choanocytes
   c. Water passes through the spongocoel and exits through the osculum
   d. Simple structural geometry limits the body size of the sponge
      (1) As the sponge grows larger, body surface area increases much more slowly than body volume (known as the surface-area-to-volume ratio); cell surface functions, such as diffusion, may not take place rapidly enough to provide for this increased mass of body cells
      (2) With the simple vase-like structure of asconoid sponges, the choanocytes that line the body cavity cannot move and filter enough water to supply food and oxygen to a thick layer of underlying cells, thereby limiting the size of the sponge
   e. Asconoid sponges are found only in class Calcarea
   f. *Leucosolenia* is a typical asconoid sponge
3. *Syconoid* sponges have a body cavity that is evaginated and folded into a series of radially arranged canals lined with choanocytes
   a. The basic shape is tubular with a single osculum, similar to asconoid sponges

Types of Sponge Structures

In each of the three structural types of sponges, the choanocytes are shown in black. Arrows indicate the direction of water flow.

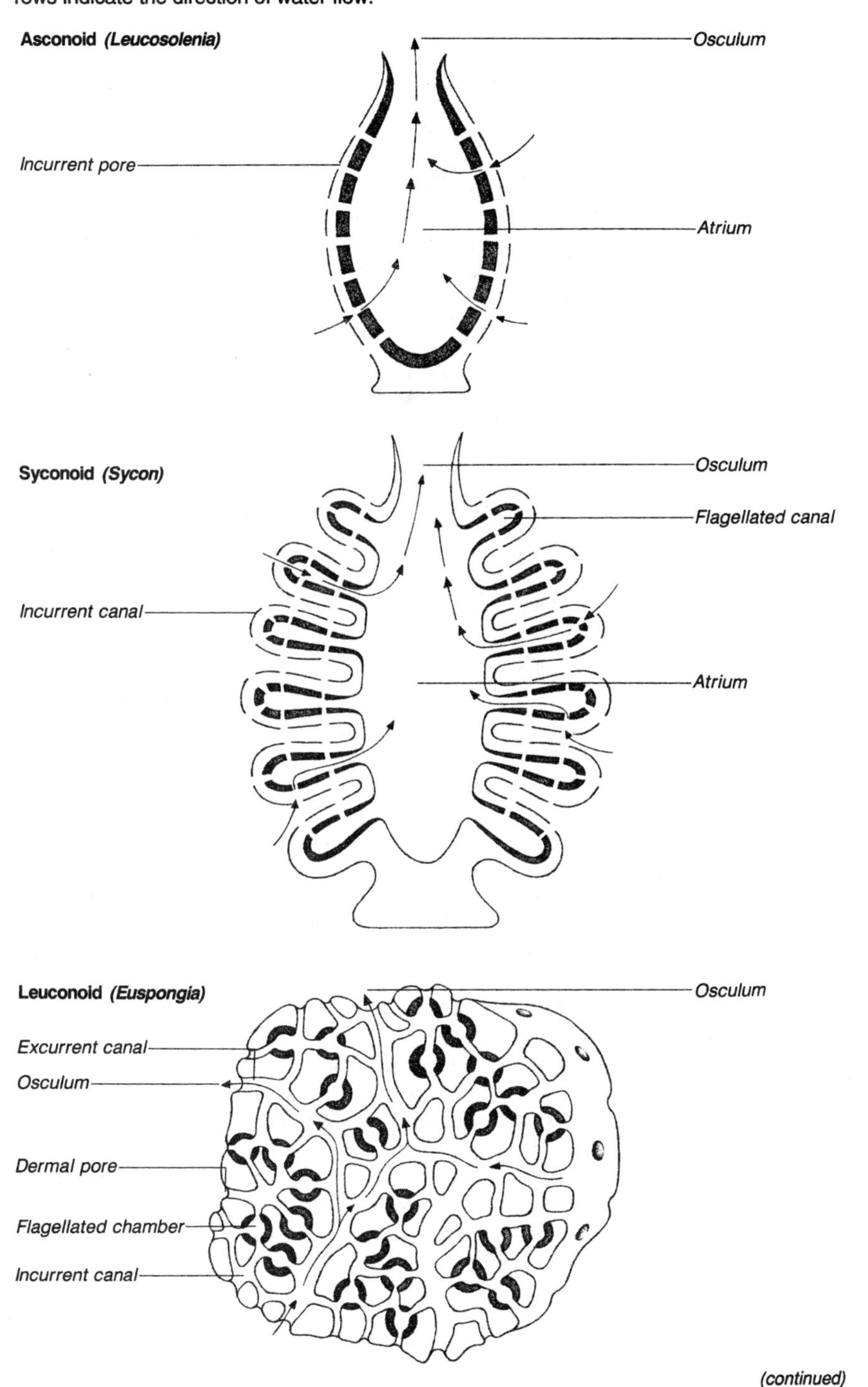

(continued)

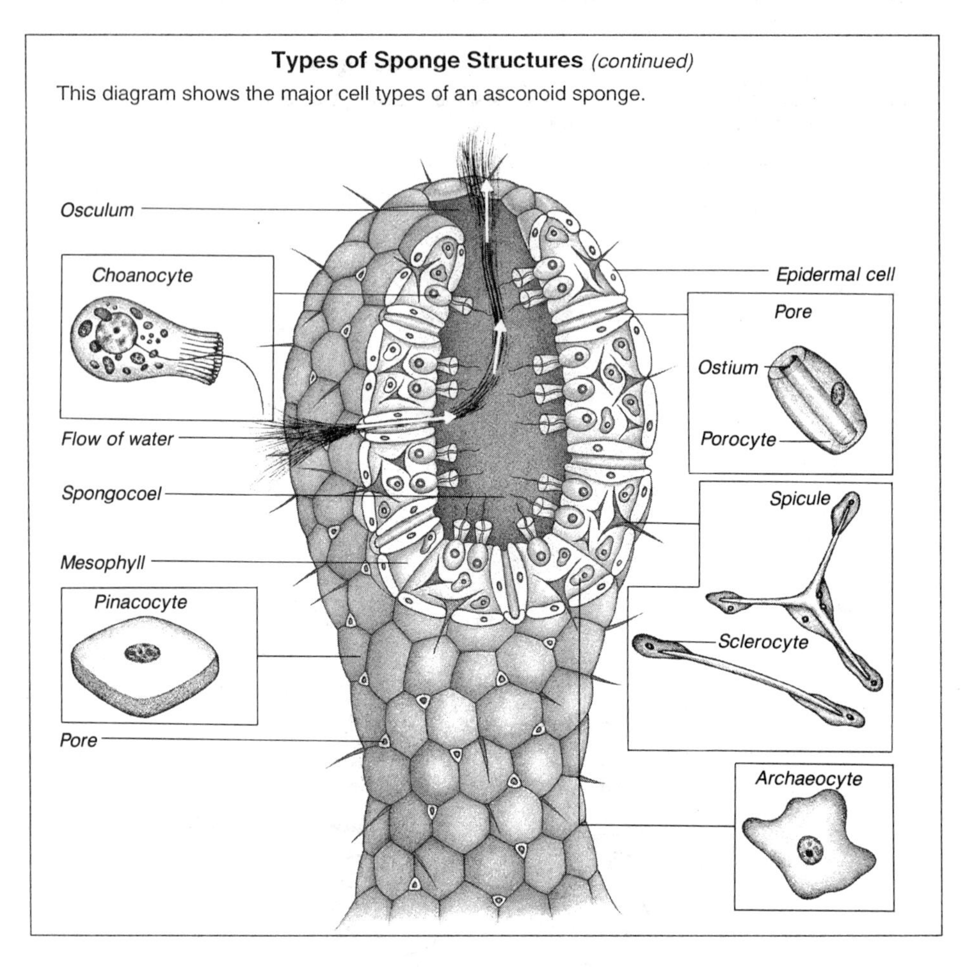

b. The body wall is thicker and more complex, but the radially arranged canal system provides a significant increase in functional surface area
c. Water enters through incurrent canals, which are larger in diameter than the microscopic incurrent pores found in asconoid sponges
d. The water passes through choanocyte-lined radial canals into the spongocoel and exits through the osculum
e. Unlike that of asconoid sponges, the syconoid spongocoel is not lined with choanocytes
f. Syconoid sponges are found in classes Calcarea and Hexactinellida
g. *Sycon* is a typical syconoid sponge

4. *Leuconoid* sponges have the most complex body type, with many branched canals and chambers lined with choanocytes
   a. Most leuconoid sponges are asymmetrical with many oscula on the surface
   b. Water enters through a branching system of incurrent canals and reaches rounded chambers lined with choanocytes
   c. The choanocyte-lined chambers discharge water into a small excurrent canal

d. The small excurrent canals merge to form larger canals; these canals join together to form a major excurrent canal, through which water reaches an osculum
e. The presence of many choanocyte-lined chambers maximizes the surface-area-to-volume ratio, greatly increasing the water-handling efficiency of the sponge
(1) The rate of water movement and volume of water filtered in leuconoid sponges far exceeds that of asconoid and syconoid sponges
(2) Because leuconoid sponges filter water most efficiently, they grow much larger than sponges with other body types
f. Most sponges are leuconoid; they are found in all three sponge classes and are the only sponge type in classes Demospongiae and Sclerospongiae
g. The commercial bath sponge *Hippospongia* and the bright yellow sponge *Cliona* are examples of leuconoid sponges

III. Internal Structure and Physiology

A. General information

1. Sponges are multicellular organisms composed of loosely organized cells
2. A few types of specialized cells perform functions such as nutrition and reproduction, but the cells are scattered and not well organized compared with those of most other animal groups
3. Most sponge body cells are extremely versatile, retaining the ability to move and change form and function
4. Time-lapse films of sponge cells show frequent changes in shape, location, and function

B. Specialized cell types

1. Specialized cells in sponges include pinacocytes, porocytes, choanocytes, archaeocytes, sclerocytes, and the mesohyl (see *Types of Sponge Structures*)
2. *Pinacocytes* are flattened epithelial cells that line the exterior surface and all internal canals and spaces that are not covered with choanocytes
a. These cells are slightly contractile and help to maintain body shape
b. Some pinacocytes secrete a collagen-polysaccharide structure that allows the sponge to attach to the substrate; others help to regulate water flow
c. In freshwater sponges, pinacocytes assist with feeding and play an important role in osmoregulation, maintaining appropriate salt and water concentrations in internal body fluids
3. *Porocytes* are tubular pore cells that pierce the body wall
a. Porocytes are present only in asconoid sponges
b. They possess a diaphragm that can open and close the pore
c. They also can ingest and digest food
4. *Choanocytes (collar cells)* line the spongocoel of asconoid sponges and the canals and chambers of syconoid and leuconoid sponges
a. Choanocytes are the most characteristic sponge cells and also the most important to the sponge life-style
b. The flagellated choanocytes create water currents that bring in food and oxygen and eliminate wastes

(1) The exposed cell end carries a flagellum surrounded by a collar-like ring of microvilli
(2) The beating of the flagellum pulls water past the cell in a continuous flow
(3) Each cell beats independently; movement is not coordinated with that of other choanocytes

c. Choanocytes also engulf food particles through phagocytosis
(1) The collar acts as a sieve and filters food particles from the water
(2) Particles too large to enter the collar are trapped in secreted mucus and slide down the collar to the base, where they are phagocytized by the cell body

5. *Archaeocytes (amoebocytes)* are highly mobile cells that are involved in most of the life functions of the sponge
a. They phagocytize and digest food particles from the water and also can ingest and digest phagocytized food particles from the choanocytes
b. Archaeocytes store food reserves and transport the stored food around the body of the sponge
c. They secrete spicules and spongin fibers
d. They transport metabolic wastes to excurrent canals for excretion
e. Archaeocytes can differentiate and give rise to any other sponge cell type

6. *Sclerocytes* are specialized cells that accumulate calcium or silicate and use it to secrete spicules; they are found in the mesohyl

7. The *mesohyl* is an extracellular matrix that often is the chief component of the sponge body
a. It is sandwiched between two thin epidermal layers; the outer cell layer consists of pinacocytes and the inner surface is made up of choanocytes
b. A variety of cell types are found in the mesohyl, along with collagen fibers and spicules
c. The jelly-like mesohyl facilitates the movement of amoeboid cells and thus plays a role in digestion, nutrient transport, and excretion

C. Sponge physiology

1. Because sponges have no specialized organs or organ systems, they perform gas exchange and excretion by diffusion
2. The continuous current of water flowing through the sponge provides food, supplies oxygen to the cells, removes carbon dioxide and metabolic wastes, and transports gametes
3. Sponges obtain nutrition by filter feeding
a. Because all digestion is intracellular, sponges can only consume particles that can be engulfed by a single cell
b. Sponges are size-selective particle feeders, ingesting only those food items that have a diameter less than that of the incurrent pores
c. Sponges feed on plankton, microscopic algae, bacteria, and organic debris; they also can absorb dissolved nutrients, such as amino acids, directly from the water
(1) Food particles captured by archaeocytes are digested in food vacuoles within the cell
(2) Food particles captured by choanocytes are partially digested within the cell's food vacuoles and then passed by exocytosis to the meso-

hyl, where they are engulfed by archaeocytes for completion of the digestion process

(3) In both cases, mobility of cells through the mesohyl ensures transport of nutrients throughout the sponge body

D. Reproduction

1. Sponges reproduce both sexually and asexually
2. Various methods of sexual reproduction are used by sponges
   a. Most species are hermaphroditic, but produce eggs and sperm at different times
   b. Sex changes may occur only once or many times in the life of an individual sponge; these animals may begin life as either male or female
   c. Sperm are produced primarily from choanocytes, while eggs may be produced by choanocytes or archaeocytes
   d. Mature sperm are released into the water current flowing through the sponge and exit through the osculum
   e. Once in the water, sperm are taken in through the incurrent pores of neighboring sponges
   f. Sperm are captured by choanocytes and, if of the same species, are enclosed in vacuoles in a manner similar to the enclosure of food particles in food vacuoles; sperm of other species are eaten
   g. The sperm-containing choanocyte loses its collar and flagellum and migrates through the mesohyl as an amoeboid cell, transporting the sperm to the egg
   h. The fertilized eggs develop into larvae in the mesohyl
   i. The free-swimming larvae are released through excurrent canals or erupt through the body wall
   j. Several hours to a few days later, depending on the species, the larvae settle to the substrate and develop into adult sponges
3. Several modes of asexual reproduction are common in sponges
   a. Marine sponges form external buds that may detach from the parent or remain attached to form colonies
   b. Internal buds called *gemmules* form in freshwater sponges and some marine species; gemmules can withstand drying and freezing
      (1) Freshwater sponges die and disintegrate in late fall, leaving gemmules to survive the winter
      (2) Sponges also produce gemmules in unfavorable conditions, such as drought
      (3) When favorable conditions return, cells inside the gemmules emerge and develop into new sponges
4. All sponges studied to date can regenerate
   a. Sponges easily regenerate damaged or missing parts
   b. Experiments in which sponge cells are completely dissociated show that the cells recombine and reorganize into a new functional sponge
   c. Some sponges possess ***recognition proteins*** on the cell surface that are specific to that species; these proteins enable isolated sponge cells to combine only with cells of the same species

IV. Classification of Phylum Porifera

A. General information

1. The three classes of sponges are Calcarea, Hexactinellida, and Demospongiae
2. The classes differ in the type of skeleton, size, body shape, body type, and habitat

B. Class Calcarea (Calcispongiae)

1. Sponges in this class have a skeleton composed of calcium carbonate spicules
2. Most of these sponges are small (less than 10 cm high)
3. They usually are tube- or vase-shaped and may exist alone or in colonies
4. The body type may be asconoid, syconoid, or leuconoid; most are drab, but some are bright red, yellow, green, or lavender
5. All sponges in this class are marine animals and most live in shallow water
6. Common examples are *Leucosolenia*, *Leucilla*, *Grantia*, and *Sycon (Scypha)*

C. Class Hexactinellida (Hyalospongiae)

1. Sponges in this class have a skeleton composed of silica spicules; each spicule has six rays, sometimes fused into long strands
2. These sponges may grow up to 1 meter in height
3. Most of these sponges are radially symmetrical
4. The body type may be syconoid or leuconoid
5. All are marine animals, and most are found in deep water (200 to 2,000 meters)
6. Common examples are *Euplectella* (Venus's flower basket), *Staurocalyptus*, and *Rhabdocalyptus*

D. Class Demospongiae

1. Sponges in this class have a skeleton composed of silica spicules, spongin, or both
2. They range from thin (encrusting forms) that are only a few millimeters thick to large individuals that are several centimeters thick and more than a meter high
3. Many are asymmetrical, have several oscula, and are brightly colored; however, a full range of shapes and colors exist
4. All Demospongiae have a leuconoid body type
5. The class includes marine and freshwater representatives; all freshwater sponges belong to this group
6. More than 80% of all sponge species belong to the class Demospongiae; common examples are *Haliclona*, *Microciona*, *Cliona* (boring sponge), and the freshwater genera *Ephydatia* and *Spongilla*

Study Activities

1. List six basic characteristics of sponges.
2. Identify possible ancestors of sponges and justify your choices.
3. Compare and contrast asconoid, syconoid, and leuconoid sponges.
4. List four cell types and explain their functions.
5. Describe how sponges obtain nutrition, exchange gases, and excrete waste.
6. Describe sexual and asexual reproduction in sponges.
7. Discuss the basic characteristics of each of the three sponge classes.

6

Radiate Phyla: Cnidaria and Ctenophora

Objectives

After studying this chapter, the reader should be able to:
- Describe the basic characteristics of the radiate phyla.
- Compare the two body types of cnidarians.
- Discuss the functions of the specialized cell types in cnidarians.
- Describe how cnidarians perform their basic life functions.
- Describe the structure and function of a nerve net.
- Identify and characterize the four classes of cnidarians.
- Compare and contrast the basic characteristics of cnidarians and ctenophorans.

I. Basic Characteristics

A. General information

1. The radiate phyla are characterized by primary radial or biradial symmetry
   a. Animals with primary radial symmetry are radially symmetrical at all stages of the life cycle, beginning at the earliest developmental stages
   b. In contrast, animals of phylum Echinodermata (see Chapter 13, Phylum Echinodermata) are radially symmetrical as adults but have larvae with bilateral symmetry
2. The two phyla of radiate animals are Cnidaria (anemones, corals, jellyfish) and Ctenophora (comb jellies)
3. The radiate phyla are considered to be at the tissue level of organization
   a. Cnidarians differ from most metazoans in that they are diploblastic; body tissues are derived from only two of the three embryonic germ layers (endoderm and ectoderm; mesoderm is lacking)
   b. Their organs, therefore, are relatively simple in structure and function
4. An important advance in the radiata is the presence of a *coelenteron,* or ***gastrovascular cavity;*** this central cavity has just one opening that functions in digestion, respiration, and circulation

B. Ecologic relationships

1. The radiata are entirely aquatic; most are marine species with a few freshwater representatives
2. They are either sessile or *planktonic* (free-floating), with only limited swimming abilities

II. Basic Characteristics of Phylum Cnidaria

A. General information

1. Animals in the phylum Cnidaria are radially or biradially symmetrical
   a. The mouth, surrounded by tentacles, is located at one end of the central axis
   b. The gastrovascular cavity may be bag-like or branched; it has only one opening, which serves as both mouth and anus
2. This group is characterized by specialized cells called *cnidocytes*, which contain stinging organelles called *nematocysts*
3. Cnidarians exhibit ***polymorphism,*** that is, they have two basic body types with many variations possible, especially in colonial (group) forms
   a. The *polyp* (or hydroid form) is adapted for a sessile (attached) or sedentary (stationary, inactive) life-style
   b. The *medusa* (or jellyfish form) is adapted for floating or swimming
   c. Although superficially different in appearance, both the polyp and medusa forms have the same basic body plan (see *Comparison of Polyp and Medusa Body Type*)
   d. Many individuals are brightly colored, including orange, pink, purple, and blue varieties

B. Ecologic relationships

1. About 9,000 living species of cnidarians have been identified
2. Cnidarians are carnivorous
   a. Small forms, such as hydras and corals, capture plankton
   b. Jellyfish and anemones prey on larger invertebrates and small fishes
3. Symbiotic associations between cnidarians and other animals are common

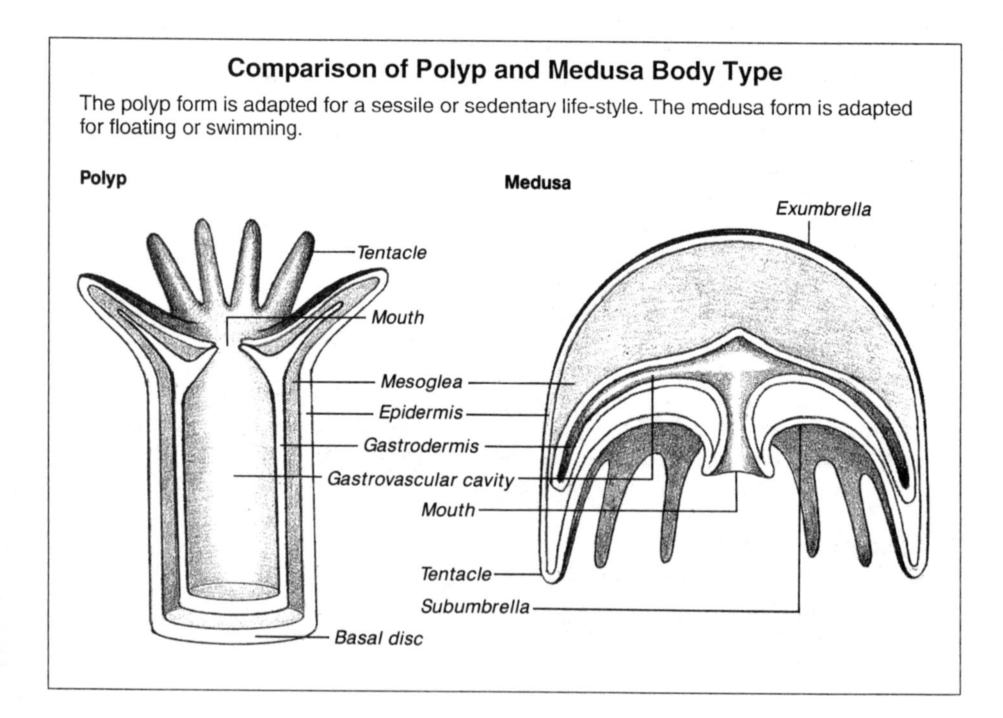

Comparison of Polyp and Medusa Body Type

The polyp form is adapted for a sessile or sedentary life-style. The medusa form is adapted for floating or swimming.

a. Hydroids and anemones may live on the shells of molluscs (see Chapter 9, Phylum Mollusca), hermit crabs, and other crustaceans (see Chapter 11, Phylum Arthropoda); in this mutualistic relationship, the polyp obtains transportation and greater availability of food resources, while the host receives camouflage and protection from the cnidarian's stinging tentacles
b. The tentacles of anemones are home to anemone fishes, which receive protection from predation
c. Several marine and freshwater cnidarians have a close mutualistic relationship with unicellular algae living inside their body tissues
(1) The algae, known collectively as ***zooxanthellae,*** receive protection and exposure to sunlight and have access to the nitrogenous waste products of their hosts
(2) The cnidarians receive organic compounds produced by photosynthesis

4. Reef-building corals are the source of one of the most productive ecosystems on earth; the coral branches provide shelter and attachment surfaces for a wide variety of plants and animals

C. Evolutionary relationships

1. The cnidarians have one of the longest fossil histories among the metazoa, stretching back as far as 700 million years
2. Current theories on the ancestry of the cnidarians suggest that they developed from a colonial flagellated protozoan
a. A spherical flagellated colony gave rise to a hollow multicellular ancestor, which evolved into a radially symmetrical, multicellular organism with two embryonic germ layers (as found in present-day cnidarians)
b. This theory is supported by the presence of radial symmetry in cnidarians
c. This same ancestral line may have given rise to all metazoan animals

III. Cnidarian Form and Function

A. General information

1. Cnidarians are radially or biradially symmetrical
2. The mouth, surrounded by tentacles, is located at one end of the central axis
3. The central body cavity, or gastrovascular cavity, functions in nutrition and excretion
a. The cavity may be bag-like or branched
b. It has only one opening, which serves as both mouth and anus
4. Cnidarians have a three-layered body plan
a. The outer body surface consists of a layer of cells called the epidermis
b. Another layer of cells, called the gastrodermis, lines the gastrovascular cavity
c. The mesoglea, a jelly-like layer, is sandwiched between these two layers

B. Body types of cnidarians

1. Cnidarians have two basic body types
a. The polyp (hydroid form) is adapted for a sessile or sedentary life-style
(1) The body is tubular
(2) The ***oral*** end, with the mouth and tentacles, is directed outward

(3) The ***aboral*** end, opposite the oral end, attaches the animal to the substrate

(4) Polyps live as single individuals or in colonies

(5) Colonies of some species, such as *Physalia* or *Obelia*, include a variety of individuals; each is specialized for a particular function, such as feeding, reproduction, or defense

b. The medusa (jellyfish form) is adapted for floating or swimming

(1) The body is a concave umbrella or bell shaped

(2) The oral end is located on the concave side of the umbrella (directed downward); tentacles ring the umbrella

(3) The aboral end is convex and directed upward

(4) The medusa resembles an inverted polyp (see *Comparison of Polyp and Medusa Body Type,* page 42)

c. Both the polyp and medusa have the three-layered body plan characteristic of cnidarians

(1) The jelly-like layer of mesoglea is much thicker in the medusa, comprising most of the body mass

(2) This mass of mesoglea gives the medusa its common name—jellyfish

2. Colonial hydroids (polyps), found in classes Hydrozoa and Anthozoa, commonly are polymorphic

a. Hydrozoan colonies contain individuals of several different functional types (called *zooids*)

(1) For example, gastrozooids are specialized for feeding and digestion; gonozooids function in reproduction; and dactylozooids are specialized for defense

(2) These specialized polyps are highly modified and quite different in ***morphology*** (body structure) from each other

(3) Common examples of hydrozoan colonies are the *Obelia*, *Hydractinia*, and hydrocorals (polyps with a calcified skeleton similar to reef-building corals) of the genus *Millepora* (sometimes called fire corals because of their powerful sting)

b. Most colonial hydrozoans consist of groups of attached polyps; however, the order Siphonophora includes large colonies of up to 1,000 medusoid and polypoid individuals

(1) The medusoid individuals are adapted as floats or swimming bells

(2) Specialized polyps are suspended from the float

(3) A well-known example is *Physalia* (the Portuguese man-of-war)

c. Anthozoan colonies are constructed around a main supporting stem to which the remainder of the colony is attached

(1) Two common zooids are autozooids (specialized for feeding) and siphonozooids (which produce a current of water through the colony)

(2) Colonial anthozoans include gorgonians (such as sea fans), *Renilla* (sea panseys), and *Ptilosarcus* (sea pens)

IV. Internal Structure and Physiology of Cnidarians

A. General information

1. Cnidarians have various types of skeletal support
2. Some cnidarians have soft or flexible support structures

a. Polyps usually are supported by a hydrostatic skeleton (in which the water-filled gastrovascular cavity is pressurized by the muscular body wall)
b. Medusae have a thick mesoglea that often is interwoven with fibers for additional support
3. Other cnidarians have hard skeletons
a. The skeleton can be horny or woodlike, such as that found in gorgonians and sea pens
(1) Calcareous sclerites (reminiscent of the spicules found in sponge skeletons) support many gorgonians; the sclerites may be fused into a solid framework similar to that found in true corals
(2) The bright colors of gorgonians are due to the presence of these sclerites (for example, the red coral used in jewelry is actually a gorgonian with fused red sclerites)
b. Thick calcareous skeletons are found in the true (stony) corals, a group that includes the massive reef-building coral colonies

B. Specialized cell types
1. Cnidarians contain various specialized cell types (see *Cross Section of a Hydroid Body Wall,* page 46)
2. The epidermal layer contains several types of cells
a. *Cnidocytes* secrete the stinging organelles called nematocysts
(1) Nematocysts serve a variety of functions, such as prey capture, defense, locomotion, and attachment
(2) They consist of a tiny capsule containing a hollow, looped thread or tube
(a) The end of the capsule is covered by an *operculum* (hinged lid)
(b) The nematocyst thread may bear poisonous barbs or spines; others are adhesive
(3) Nematocysts may contain powerful ***neurotoxins*** (poisons that inhibit the nervous system), which are capable of immobilizing large active prey (such as fish) and seriously injuring or killing humans
(4) Each nematocyst can be fired only once
(a) Nematocysts fire when chemically stimulated (such as by the presence of various sugars or amino acids) or when they are touched
(b) Cnidarians have only minimal control over the firing of nematocysts; firing can occur even if the animal is dead
(5) More than 20 different types of nematocysts are found in this phylum; nematocysts are used to identify different species
b. *Epitheliomuscular cells* are contractile epidermal cells that provide protection and support; contractions of these cells help to move the tentacles
c. *Sensory cells* are abundant around the mouth and tentacles; they are chemical and tactile receptors
3. The gastrodermis contains specialized cells called *nutritive-muscular cells*
a. These primitive muscle cells form the longitudinal and circular muscle layers
b. In polyps, alternate contraction of these two layers facilitates locomotion
c. These muscle layers also maintain water pressure in the gastrovascular cavity to support the hydrostatic skeleton (which also plays a role in locomotion)
d. Nutritive-muscular cells are flagellated; the beating flagella circulate food and fluids in the gastrovascular cavity

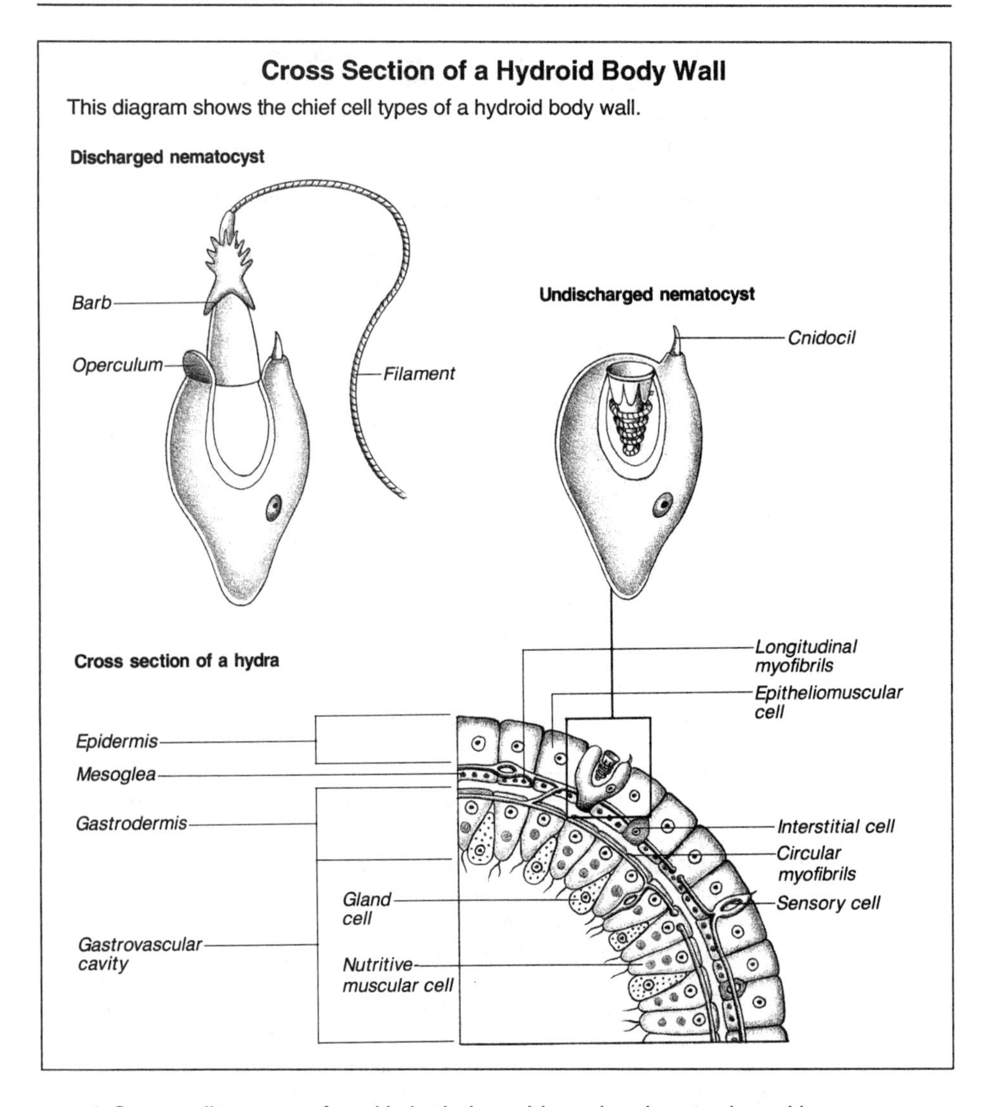

4. Some cell types are found in both the epidermal and gastrodermal layers
    a. *Interstitial cells* are undifferentiated cells that give rise to most of the other cell types (with the exception of epitheliomuscular cells, which are self-reproducing); the interstitial cells are similar to the archaeocytes of sponges
    b. *Gland cells* around the mouth and base secrete mucus or adhesive material; those in the gastrodermis secrete digestive enzymes
    c. *Nerve cells* are found throughout the epidermis and gastrodermis; cnidarians have a loosely organized nervous system of interconnected nerve cells, called a *nerve net*

C. Nutrition

1. All cnidarians are carnivorous; nematocysts on feeding tentacles paralyze or entangle prey, which then is carried to the mouth

2. Preliminary digestion is extracellular and begins in the gastrovascular cavity; digestive enzymes are secreted by the gland cells of the gastrodermis
3. After this initial breakdown, the food particles are phagocytized by the nutritive-muscular cells of the gastrodermis, where intracellular digestion occurs in food vacuoles
4. Waste products return to the gastrovascular cavity and are expelled through the mouth

D. Reproduction

1. Reproduction in cnidarians is characterized by ***alternation of generations,*** in which an asexually reproducing polyp stage alternates with a sexually reproducing medusa stage
    a. Although alternation of generations is the basic pattern, some groups have developed variations on this theme
    b. Polyps of class anthozoa are capable of both asexual and sexual reproduction; the medusa stage is absent
    c. As with many sessile organisms, cnidarians also have a dispersal stage consisting of free-swimming larvae
2. Most cnidarians are ***dioecious*** (male and female sexes are in separate individuals)
    a. During sexual reproduction, gametes develop from interstitial cells and accumulate in specific locations in the epidermis or gastrodermis
    b. Gametes are released directly into the water; fertilization occurs externally
    c. Development takes place as the embryos float freely in the plankton
        (1) Embryos develop into *planula* larvae (a slightly elongate, radially symmetrical, ciliated larval stage)
        (2) Several hours to a few days later, depending on the species, the larvae settle to the substrate and develop into young polyps
3. Asexual reproduction is common, especially in polyp forms
    a. Polyps form external buds, which may detach from the parent or remain attached to form colonies
    b. Asexual reproduction in anemones occurs by ***pedal laceration;*** small pieces of the pedal disc (attachment base) break off as the animal moves, and each piece develops into a small anemone
    c. Cnidarians easily regenerate damaged or missing parts

E. Circulation, gas exchange, excretion, and osmoregulation

1. Cnidarians have no special organs for circulation, respiration, or excretion
2. Nutrients circulate through the gastrovascular cavity
3. Gas exchange and excretion of nitrogenous wastes occurs by diffusion through the internal and external body walls

F. Nervous system and sense organs

1. Cnidarians have a diffuse, noncentralized nervous system known as a nerve net; this phylum is the first in which muscular contraction primarily is under nervous control
2. The nerve net consists of two main networks of interlocking nerve cells; one located in the epidermal layer and one in the gastrodermal layer

a. Most nerve conduction is nonpolar, that is, the impulse is sent in both directions at once; a stimulus may spread the nerve impulses in every direction throughout the body
b. In medusae, the epidermal nerve net is concentrated into two nerve rings near the bell perimeter; these connect with fibers innervating the tentacles, muscles, and sense organs to facilitate movement
c. Remnants of nerve nets are found in all higher animal groups, including humans (where it is found as nerve complexes in the digestive system musculature)

3. Because muscle contractions occur slowly, sessile members of this phylum appear motionless most of the time
   a. Continuous muscular activity does occur; cnidarian behaviors simply take place on a slow time scale
   b. Cnidarians can respond quickly to environmental stimuli, such as in feeding or when disturbed
4. The nervous systems and sense organs of medusae are more complex than those of polyps
   a. Statocysts are distributed around the bell perimeter
      (1) A medusa rights itself by contracting muscles on the side opposite the statocyst stimulation
      (2) Medusae with symbiotic zooxanthellae (such as *Cassiopeia*) use statocysts to help them orient their bodies for maximal light exposure
   b. Chemoreceptors and ocelli also are distributed around the body; some medusae have well developed eyes with a cornea, retina, and lens

V. Classification of Phylum Cnidaria

A. General information

1. The four classes of cnidarians are Hydrozoa, Scyphozoa, Cubozoa, and Anthozoa
2. The classes differ in type of skeletal support, body shape, and type of reproduction

B. Class Hydrozoa

1. Most hydrozoans are marine animals, but there are a few freshwater representatives; this class includes many colonial species, commonly called *hydroids*
2. Hydrozoans usually are small and inconspicuous but abundant in a variety of habitats; most hydroid colonies are 3 to 10 centimeters high
3. Hydrozoan polyps usually are colonial with a high degree of polymorphism
4. Both polyp and medusa stages are present in the life cycle; however, some species (such as the freshwater hydra) have no medusa stage
   a. Medusae are produced by budding within specialized polyps called gonangia
   b. Young medusae leave the colony as free-swimming organisms and produce gametes (eggs or sperm) that are shed into the water; in some species, the medusae remain attached to the colony
   c. The planula larva that results from fertilization settles to the substrate and develops into a young polyp; this polyp gives rise, by asexual budding, to a new hydroid colony

5. Hydrozoan medusae (jellyfish) usually are small, ranging from a few millimeters up to a few centimeters in diameter
   a. They can be identified by the presence of a *velum,* a shelf-like inward fold of the bell margin; when the bell contracts for swimming, the velum increases the force of the water jet
   b. The gastrovascular cavity is not a single sac as in polyps; the mouth opens into a central stomach from which four radial canals extend; the radial canals in turn connect with a ring canal that extends around the bell margin
   c. The gametes of hydrozoan medusae develop in gonads located within the epidermis beneath the radial canals
6. Common examples of the class Hydrozoa are the freshwater *Hydra*, *Millepora* (fire coral), *Obelia*, and *Craspedacusta*

C. Class Scyphozoa

1. The class Scyphozoa is characterized by large, conspicuous medusae; most common jellyfish belong to this class
2. The polyp is small and inconspicuous and may be absent in some groups
3. Most scyphozoan medusae range in size from 2 to 40 centimeters in diameter; some, such as *Cyanea* (the lion's mane), have a diameter of more than 2 meters with tentacles 60 to 70 meters long
4. Most schyphzoans are pelagic (found in the open ocean); they are distributed in oceans around the world (even in cold Arctic waters) at depths of up to 3,000 meters
5. Scyphozoan medusae have no velum; swimming occurs by bell contractions
6. Schyphzoan medusae have four long divisions of the body wall, called oral lobes, that hang beneath the umbrella; these lobes are used for food capture and ingestion
   a. The tentacles, which vary in number and length, are distributed around the bell margin
   b. Nematocysts can produce painful stings, and even cause death in humans
   c. The internal body structure is similar to that found in hydrozoan medusae, including the presence of a central stomach divided into four gastric pouches, radial canals, and a ring canal; statocysts and ocelli are distributed around the bell margin
7. Gametes develop in the gastrodermis and collect in gonads in the gastric pouches; eggs and sperm are shed through the mouth
   a. The embryo develops into a scyphistoma larva (a tiny polyp form), which attaches to the substrate
   b. The scyphistoma produces asexual buds called *strobila*, which develop into young medusae (called *ephyrae*)
   c. The ephyrae break loose, swim away, and grow into adult, sexually reproducing medusae
8. Common examples of the class Schyphozoa are *Aurelia* (moon jelly), *Pelagia*, *Cyanea* (lion's mane), and *Cassiopeia*

D. Class Cubozoa

1. Members of the class Cubozoa share many characteristics with the scyphozoans
2. Medusae are dominant in this class but differ from scyphozoan medusae in that they are smaller, box shaped, and usually colorless

3. No strobila form during the reproductive cycle; the scyphistoma larva develops into a single adult medusa
4. The sting of cubozoans is dangerous and sometimes fatal in humans; death can occur within a few minutes, a characteristic that gives the group its common name—the sea wasps
5. Cubozoan medusae are strong swimmers and powerful predators; they feed primarily on fishes
6. Examples include *Carybdea* and *Chironex*

E. Class Anthozoa

1. With more than 6,000 species, class Anthozoa is the largest class of cnidarians
2. The class is exclusively marine; representatives are found in all seas at a wide range of depths
3. Anthozoans are polyps whose tentacles resemble flower petals
4. The polyps may be individual or colonial; the medusa stage is absent in this group
5. Many species are supported by external skeletons, primarily composed of calcium carbonate
6. Anthozoans vary greatly in size; the largest animals in this class are the anemones, named for their flower-like appearance; they range from a few centimeters to 1 meter in diameter
7. The body structure of anthozoans has several unique features
    a. The gastrovascular cavity is large and divided into compartments by sheet-like extensions of the body wall called ***septae***
    b. The pharynx is a muscular extension of the mouth
        (1) Ciliated grooves extend from the mouth to the pharynx
        (2) The cilia help pump water through the gastrovascular cavity
        (3) These water currents carry in oxygen and remove wastes; they also help to maintain fluid pressure in the hydrostatic skeleton
8. This class contains several different groups
    a. The *anemones* are the largest anthozoans; these large, heavy polyps live individually or in colonies
    b. The *scleractinian* or *stony corals* are small colonial polyps that collectively secrete a massive calcium carbonate skeleton (coral reef); the gastrovascular cavities of all the polyps in the colony are interconnected
    c. The *octocorals* (which include sea fans, sea whips, and soft corals) are small colonial polyps that have, as their name implies, only eight tentacles and eight mesenteries dividing the gastrovascular cavity; as with stony corals, polyps in the colony are interconnected
9. Typical examples of the class Anthozoa are the anemones *Metridium* and *Actinia*, stony corals such as *Porites* and *Siderastraea*, and gorgonians such as *Acanthogorgia* and *Gorgonia*

VI. Basic Characteristics of Phylum Ctenophora

A. General information

1. Members of this phylum are transparent animals that drift in open ocean planktonic environments
2. Commonly called comb jellies or sea walnuts, a few species are benthic (bottom dwellers)

3. They have a variety of body shapes, ranging from globes about 1 centimeter in diameter to long ribbons up to 1 meter long
4. Ctenophores exhibit ***bioluminescence;*** they emit light by the breakdown of specialized proteins
5. Except for one species, ctenophores do not have nematocysts
6. Examples of ctenophores are *Pleurobranchia*, *Cestum* (Venus's girdle), and *Mnemiopsis*

B. Ecologic relationships

1. This small phylum contains approximately 100 described species, all of which are marine animals
   a. Most ctenophores are free swimming
   b. They are present in all seas but primarily inhabit tropical waters
   c. They are found in shallow surface water and at depths of up to 3,000 meters
2. In certain locations and at certain times of the year, ctenophores comprise a large percentage of the planktonic community; thus, they are important in oceanic food chains
3. Ctenophores are predatory carnivores, capturing their prey by means of adhesive secreting cells on their tentacles

VII. Ctenophoran Form and Function

A. General information

1. Ctenophores are biradially symmetrical (see *Biradial Symmetry in Phylum Ctenophora,* page 52)
2. They do not have a hard skeleton and have no special organs for respiration, excretion, or circulation
3. On the body surface, ctenophores have eight longitudinal bands of cilia called *comb rows;* these ciliated bands beat in unison to propel the animal forward
4. The body has two layers that are separated by a cellular mesenchyme
5. The gastrovascular cavity is branched, and the gut ends in two small anal pores
6. Statocysts and other sensory cells are located in the epidermis

B. Nutrition

1. Ctenophores are predatory carnivores that capture their prey with special adhesive-secreting cells on their tentacles, called *colloblasts*
   a. Small prey organisms adhere to the tentacles
   b. As the tentacles fill with food, they contract and deposit the food on the mouth
2. Preliminary digestion is extracellular and occurs in the pharynx; after this initial breakdown, food particles are phagocytized by gastrodermal cells, where intracellular digestion occurs in food vacuoles
3. Waste products are expelled through small pores in the aboral end

C. Reproduction

1. Ctenophores reproduce both sexually and asexually
2. Most ctenophores are simultaneous hermaphrodites and thus are capable of self-fertilization during sexual reproduction
   a. Gametes develop in the gastrodermis and accumulate in the digestive canals
   b. They are released directly into the water, where fertilization occurs

Biradial Symmetry in Phylum Ctenophora

Ctenophorans are named for their eight rows of ciliary combs. Biradial symmetry is apparent from the pair of long, retractable tentacles.

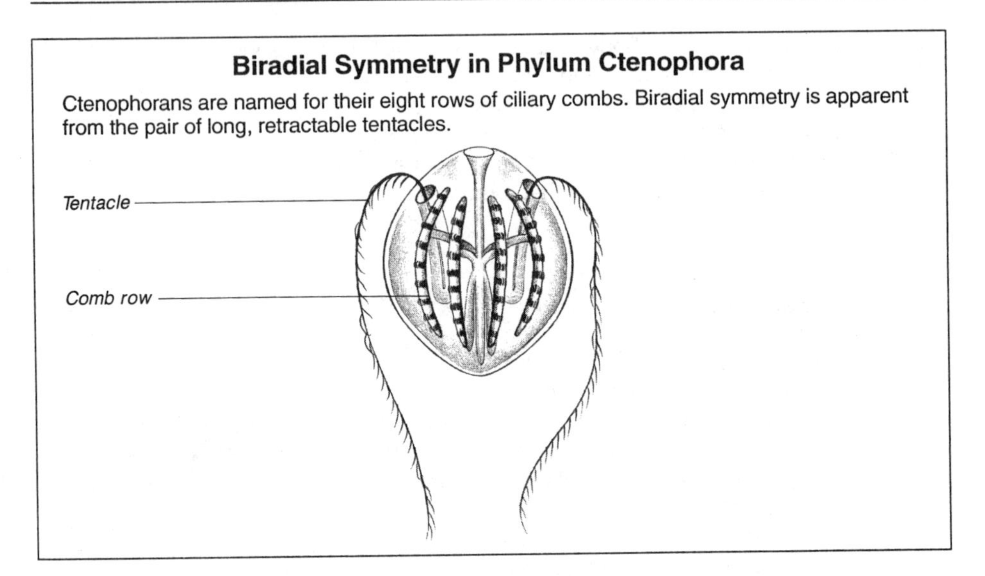

c. The embryos develop into adults while floating freely in the plankton
   (1) Embryos develop into *cydippid* larvae, which resemble small adults
   (2) The larvae gradually grow to adult size
3. Benthic ctenophores use a form of asexual reproduction
   a. The process is similar to pedal laceration in anemones
   b. Small pieces that break off as the animal crawls along develop into new ctenophores
4. Ctenophores can easily regenerate damaged or missing parts

D. Circulation, gas exchange, excretion, and osmoregulation
1. Ctenophores have no special organs for circulation, respiration, or excretion; these functions take place in a fashion similar to that of the cnidarians
2. Movement of water over the body surface is facilitated by beating of the cilia-covered comb plates

E. Nervous system and sense organs
1. Ctenophores have a nerve net system similar to that of the cnidarians
2. The main network of interlocking nerve cells is located beneath each comb plate
3. The sense organs are similar to those found in medusoid cnidarians

Study Activities

1. List three basic characteristics of the radiate phyla.
2. Compare the polyp and medusa body type of cnidarians.
3. Select three cell types and explain their functions.
4. Describe the structure and function of a nerve net.
5. Describe how cnidarians obtain nutrition, perform gas exchange, and excrete wastes.
6. Describe sexual and asexual reproduction in cnidarians.
7. Discuss the basic characteristics of each of the four cnidarian classes.
8. Compare and contrast the characteristics of cnidarians and ctenophorans.

7

Acoelomate Phyla: Platyhelminthes and Nemertea

Objectives

After studying this chapter, the reader should be able to:
- Describe the basic characteristics of the acoelomate phyla.
- Explain current hypotheses concerning the evolutionary ancestry of flatworms.
- Describe how flatworms perform their basic life functions.
- Identify and characterize the classes of phylum Platyhelminthes.
- Compare and contrast the basic characteristics of flatworms and ribbon worms.

I. Basic Characteristics

A. General information

1. The acoelomate phyla are characterized by the absence of a body cavity surrounding the gut
2. Acoelomate animals are triploblastic
3. Because of the presence of mesoderm, from which most internal organs are derived, acoelomate organs are more specialized in structure and function than those in the radiate phyla; acoelomates are at the organ-system level of organization
4. Acoelomates are dorsoventrally flattened, usually unsegmented worms
5. The acoelomate phyla are the most primitive of the bilaterally symmetrical animals; two representative acoelomate phyla are Platyhelminthes (flatworms) and Nemertea (ribbon worms)

B. Ecologic relationships

1. Acoelomates include both parasitic and free-living forms
   a. The parasitic forms can be either endoparasites or ectoparasites; they infect a variety of invertebrate and vertebrate hosts, including humans
   b. The free-living animals are primarily aquatic but also can be found in moist terrestrial habitats
2. Most aquatic acoelomates are benthic (bottom-dwelling); they frequently are found in crevices, under rocks and logs, and in the tiny spaces between sediment particles

II. Basic Characteristics of Phylum Platyhelminthes

A. General information

1. The flatworm body is bilaterally symmetrical and dorsoventrally flattened, with some cephalization
2. Most flatworms are unsegmented (with the exception of class Cestoda, the tapeworms)
3. These animals are triploblastic and acoelomate; spaces between the organs are filled with *parenchyma,* a mass of spongy cells, derived from mesoderm
4. Most species have an incomplete digestive tract, with the mouth as the only opening into the tract; tapeworms do not have a digestive tract
5. This group is characterized by specialized osmoregulatory structures known as *protonephridia* or *flame cells*

B. Ecologic relationships

1. About 20,000 living species of flatworms have been identified
2. Most species are parasitic; free-living forms are found in the class Turbellaria
   a. Parasitic species are primarily endoparasites (classes Trematoda and Cestoda); most species in class Monogenea are ectoparasites
   b. Free-living species predominantly are aquatic bottom dwellers and are found in both marine and freshwater environments; a few inhabit moist terrestrial areas

C. Evolutionary relationships

1. There are several, widely varying hypotheses about the origin of flatworms; some have been rejected based on recent evidence
2. One hypothesis suggests that flatworms arose from ciliates; this hypothesis no longer is generally accepted partly due to the cellular nature of the acoelomate endodermis
3. An alternate hypothesis suggests that flatworms arose from ctenophores, which became flattened and adopted a benthic life-style; this hypothesis also has been rejected because of the complex structural changes that would have been required
4. Two additional hypotheses on flatworm origins have been proposed, but neither has been conclusively validated
   a. The first hypothesis suggests that both flatworms and cnidarians arose from a common *planuloid* ancestor (a flattened, free-swimming, ciliated larval form)
      (1) Some descendants became free-floating or sessile; these developed into the cnidarians
      (2) Others developed bilateral symmetry and adopted a benthic life-style; these evolved into flatworms
   b. The second hypothesis suggests that flatworms are degenerate descendants of an unspecified coelomate ancestor
      (1) This could have occurred through ***neoteny*** (retention of embryonic or juvenile characteristics by the adult) during embryonic development
      (2) Alternatively, acoelomate and pseudocoelomate animals may have branched off simultaneously from a common coelomate ancestral group

III. Flatworm Form and Function

A. General information

1. As the name implies, flatworms are flattened, ribbon-like organisms that are adapted for crawling and maneuvering through interstices (narrow spaces)
2. These multicellular animals have complex organ systems; the reproductive system is especially intricate
3. Flatworms do not have circulatory or respiratory systems; the gastrovascular cavity functions in both digestion and circulation
4. No body cavity surrounds the gut; the space is filled with a cellular parenchyma
5. Flatworms have a well-developed, coordinated locomotion system
6. Compared with previously discussed phyla, flatworms display some of the most important advances found in the animal kingdom
    a. They are the first phylum to be triploblastic, and they have bilateral symmetry; these two characteristics are closely associated with the evolution of complex organ systems
    b. Flatworms are cephalized and have specialized divisions within the nervous system for sensory input, motor control, and nerve function integration
    c. These advances (triploblasty, bilateral symmetry, and cephalization) permit a more active life-style and the exploitation of new habitats

B. Body types of flatworms

1. Most individuals are small and have highly flattened bodies (see *Turbellarian Body Form and Structures,* page 56)
2. The flatworm body wall is composed of circular, longitudinal, and diagonal muscle layers; additional muscle fibers cross the body through the parenchyma layer
3. Free-living flatworms have a ciliated epidermis that functions in locomotion; parasitic forms have a highly resistant epidermis called a *tegument*
    a. The tegument is nonciliated and formed from cytoplasmic extensions of parenchyma cells
    b. It provides an amazing degree of protection, especially in the case of intestinal parasites (such as tapeworms), which reside in an environment of digestive enzymes yet are not themselves digested
4. The epithelium contains several types of gland cells
    a. They secrete mucus, which aids in locomotion, prey capture, and swallowing and provides protection from desiccation (drying out)
    b. They also secrete adhesive substances, which facilitate attachment to the substrate or to a host (in parasitic forms)

IV. Internal Structure and Physiology of Flatworms

A. General information

1. The mouth is the only opening to the gastrovascular cavity; the digestive system is absent in class Cestoda (tapeworms)
2. Osmoregulation and a limited amount of excretion are carried out by specialized structures called *protonephridia*
3. Flatworms are cephalized; the nervous system is composed of longitudinal nerve cords with ladder-like cross-connections

Turbellarian Body Form and Structures

The illustrations below show the body form of an adult turbellarian, *Dugesia,* and its digestive, nervous, and reproductive systems.

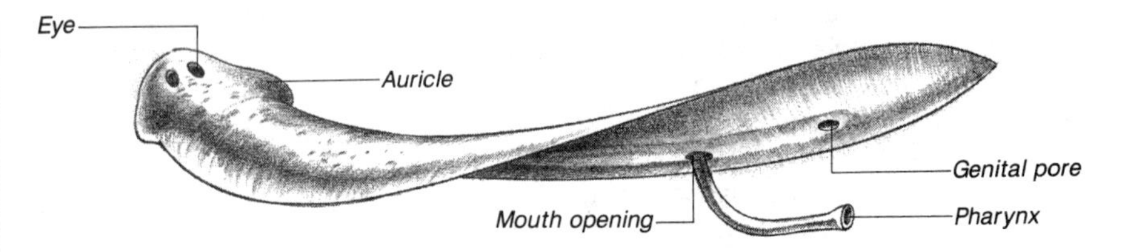

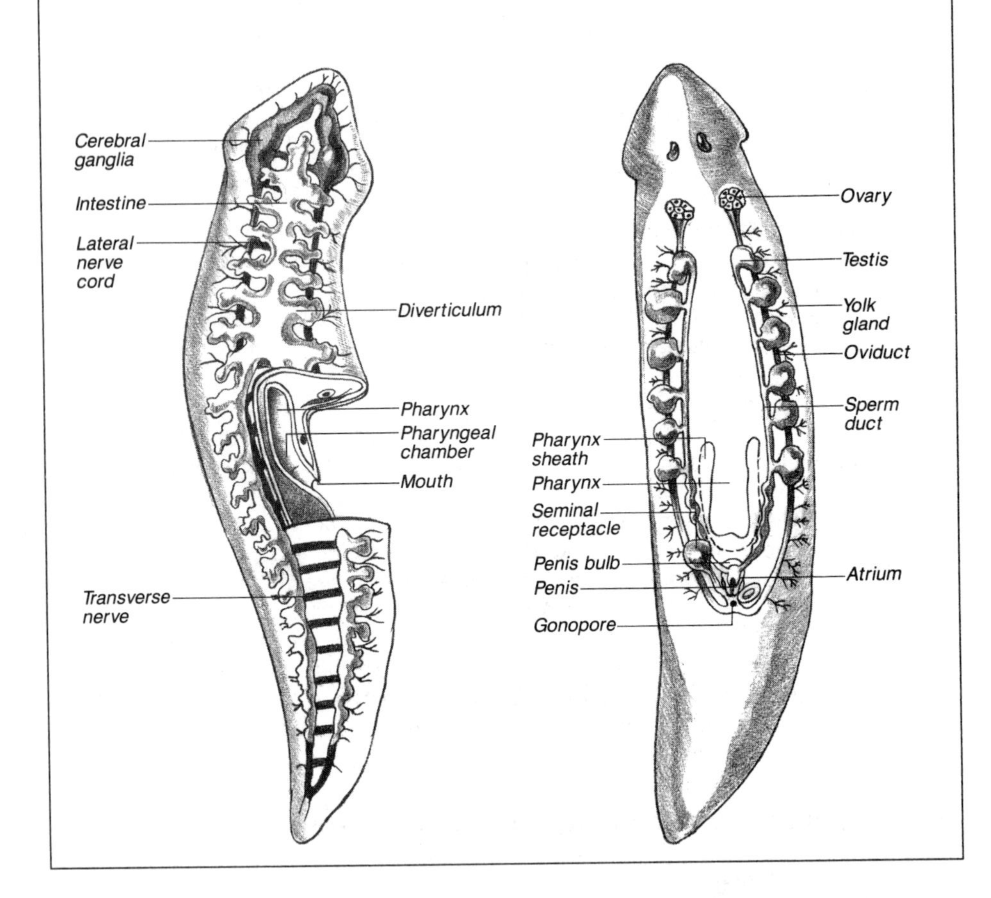

B. Locomotion

1. Benthic turbellarians move on their ventral surface; they glide over the substrate on a layer of mucus, powered by the cilia-covered epidermis
2. Aquatic forms swim by ciliary action; the motion is assisted by muscular contractions of the lateral body wall
3. Both endoparasites and ectoparasites lack a ciliated epithelium

a. Adults have limited mobility accomplished by muscular action; some adults are carried by the host's circulatory system
b. In contrast, many larval forms are actively mobile and move by ciliary action
4. Many parasitic forms have adhesive organs for attachment to the host

C. Nutrition

1. Free-living flatworms (class Turbellaria) are carnivorous predators or scavengers with a broad diet
   a. Prey include most invertebrates that are small enough to be captured and consumed; typical prey are small crustaceans, nematodes, insects, and rotifers
   b. Food is ingested whole or fragmented by a muscular protrusible (extendable) pharynx, which secretes digestive enzymes to break down the prey body
      (1) Digestion is similar to that of cnidarians; initial extracellular digestion within the gastrovascular cavity is followed by intracellular digestion by cells lining the cavity
      (2) Because the gut has only one opening, undigested material is ejected through the mouth
      (3) The intestine may be simple or highly branched (see *Turbellarian Body Form and Structures*)
2. Adult flukes (classes Monogenea and Trematoda) feed primarily on host tissues and body fluids; some feed on digested material within the host gut
   a. Food is pumped in by the muscular pharynx; some flukes secrete digestive enzymes that break down the tissue before ingestion
   b. In some cases, nutrient molecules are absorbed across the tegument
   c. Digestion is similar to that of turbellarians (a combination of extracellular and intracellular digestion)
3. Tapeworms (class Cestoda) have no mouth or digestive tract
   a. All nutrients are absorbed across the tegument
   b. Research suggests that tapeworms can absorb only small, predigested molecules

D. Reproduction

1. Flatworms can reproduce asexually
   a. Most turbellarians reproduce asexually by transverse fission or ***fragmentation***
   b. In flukes, asexual reproduction is an integral part of the life cycle and takes place in a snail intermediate host
   c. Tapeworms reproduce asexually (by budding) in their intermediate host
2. Flatworms also reproduce sexually
   a. Although most flatworms are hermaphroditic, self-fertilization does not usually occur; the exception is the tapeworm, in which self-fertilization (between segments of the same worm) is common
   b. Sexual reproductive systems are complex in anatomy and organization
      (1) The typical male reproductive organs consist of single, paired, or multiple testes (see *Turbellarian Body Form and Structures*)
         (a) The testes are drained by collecting tubules, which lead into one or more sperm ducts
         (b) The sperm ducts join together to form an ejaculation duct, which exits through the penis

(c) Sperm usually are stored before copulation in a seminal vesicle
(d) Seminal fluid from prostatic glands empties into the seminal vesicle
(e) The seminal vesicle is part of a muscular chamber called the *male atrium*, which contains the penis

(2) The typical female reproductive system has single, paired, or multiple ovaries and yolk glands
(a) The single pair of oviducts empties into the female atrium; in trematodes, the oviduct is enlarged into a coiled uterus in which many fertilized eggs are stored
(b) The female atrium also may contain specialized chambers for receipt of sperm (copulatory bursa) and postcopulatory sperm storage (seminal receptacle)

3. Parasitic flatworm species have a very high reproductive rate
4. Turbellarians can easily regenerate damaged or missing parts

E. Circulation, gas exchange, excretion, and osmoregulation

1. Flatworms have no special gas exchange or circulatory structures
   a. Gas exchange occurs by diffusion across the body wall
   b. Nutrients are distributed through the digestive system and by diffusion (facilitated by muscular movements)
2. Endoparasites are anaerobic (not dependent on oxygen for respiration) and are either facultative or obligate anaerobes; obligate (required) anaerobes cannot use and may be poisoned by oxygen while facultative (optional) anaerobes can function with or without the presence of oxygen
3. Osmoregulation and a limited amount of excretion are performed by protonephridia
   a. Compared with marine forms, freshwater turbellarians have more protonephridia with more complex tubule systems
   b. In turbellarians and flukes, only a small amount of ***ammonia*** (a nitrogenous metabolic waste product) is released by protonephridia; most ammonia is excreted by diffusion across the body wall
   c. Little is known about the role of protonephridia in tapeworms, but they may function in both excretion of metabolic wastes and osmoregulation

F. Nervous system and sense organs

1. The flatworm nervous system is composed of pairs of longitudinal nerve cords with ladder-like cross-connections (see *Turbellarian Body Form and Structures,* page 56)
2. The brain is a mass of ganglion cells located at the anterior end of the animal
3. Most flatworms have sensory, motor, and association neurons; they are the most primitive phylum to demonstrate this feature
4. Flatworms have several types of specialized sense organs
   a. *Chemoreceptors* are used for locating food
      (1) They are highly sensitive to gradients of dissolved organic molecules
      (2) Some freshwater turbellarians, such as the familiar planaria, have chemoreceptors in ear-like lobes on the sides of the head, called *auricles*
      (3) Others species have chemoreceptors in ciliated pits or grooves, on tentacles, or distributed across the anterior end of the body

b. *Ocelli, statocysts,* and ***rheoreceptors*** (which sense water currents) are distributed over the body surface of turbellarians
c. *Tactile receptors* are found all over the body of turbellarians, on the suckers of flukes, and on the scolex of tapeworms

V. Classification of Phylum Platyhelminthes

A. General information

1. The four classes of flatworms are Turbellaria, Monogenea, Trematoda, and Cestoda
2. The classes differ in life-style, body shape, and internal anatomy

B. Class Turbellaria

1. Turbellarians primarily are free-living, aquatic flatworms
   a. Aquatic species are benthic
   b. A few species are found in moist terrestrial habitats
2. About 3,000 living species have been identified
3. Turbellarians range in length from less than 5 millimeters to more than 60 centimeters; the smallest representatives are common in the ***interstitial spaces*** of subsurface sediments
4. Most species have a ciliated epidermis, and they glide over the substrate with a combination of ciliary and muscular action
5. In contrast to the other flatworm classes, turbellarians have relatively simple life cycles
   a. Asexual reproduction is by transverse fission
   b. The adults are simultaneous hermaphrodites; cross-fertilization is the general rule in sexual reproduction
   c. With the exception of a few species, zygotes undergo direct development (there is no larval stage)
   d. Some freshwater turbellarians produce two different types of eggs
      (1) Summer eggs are thin-shelled and hatch rapidly
      (2) Winter eggs are thick and resistant to adverse environmental conditions
6. Common examples of turbellarians are the freshwater planarians *Dugesia* and *Polycelis*, *Acanthiella* (interstitial flatworms with internal skeletons), and the marine flatworm *Prostheceraeus*

C. Class Monogenea

1. The class Monogenea is exclusively parasitic
   a. Most monogeneans are ectoparasites of aquatic vertebrates
   b. They frequently are found attached to the gills or external surfaces of fish, where they feed on host tissues and fluids
2. Monogeneans, commonly called *monogenetic flukes*, consist of about 1,100 living species
3. The body is covered with a nonciliated tegument
4. Members of this class have a large posterior attachment organ with suckers and hooks, called an *opisthaptor*; this organ differentiates them from the parasitic flukes of class Trematoda
5. Monogenetic flukes have a simple life cycle that involves only one host; there is no intermediate host

a. Each egg develops into a single adult worm, giving the class its name (monogenea—one generation)

b. Eggs hatch into free-swimming *oncomiracidium* larvae; on contacting a suitable host, the larvae metamorphose into adult flukes

6. Examples of monogeneans are *Dactylogyrus* (an economically important ectoparasite of freshwater fishes) and *Polystoma*

D. Class Trematoda

1. The class Trematoda is exclusively parasitic
   a. Adults usually are endoparasites of vertebrates, and the intermediate host usually is a mollusc or a snail
   b. These parasites feed on host tissues, fluids (especially blood), cell fragments, or mucus
   c. Many species cause serious diseases in humans and domestic livestock
2. Trematodes, commonly called *flukes*, have a flattened, leaf-like shape; they are structurally similar to turbellarians
3. About 11,000 living species have been described
4. The body is covered by a nonciliated tegument that typically has two hookless suckers at the anterior end *(oral sucker)* and on the ventral surface *(acetabulum)* (see *Structure and Life Cycle of a Trematode*)
5. The digestive tract is composed of a muscular pumping pharynx leading from the mouth, a short esophagus, and usually two convoluted intestinal sacs
6. Cross-fertilizing sexual reproduction is the most common form of reproduction
   a. As with most parasites, large numbers of offspring are produced
   b. Egg production may be up to 100,000 times greater than that of free-living turbellarians

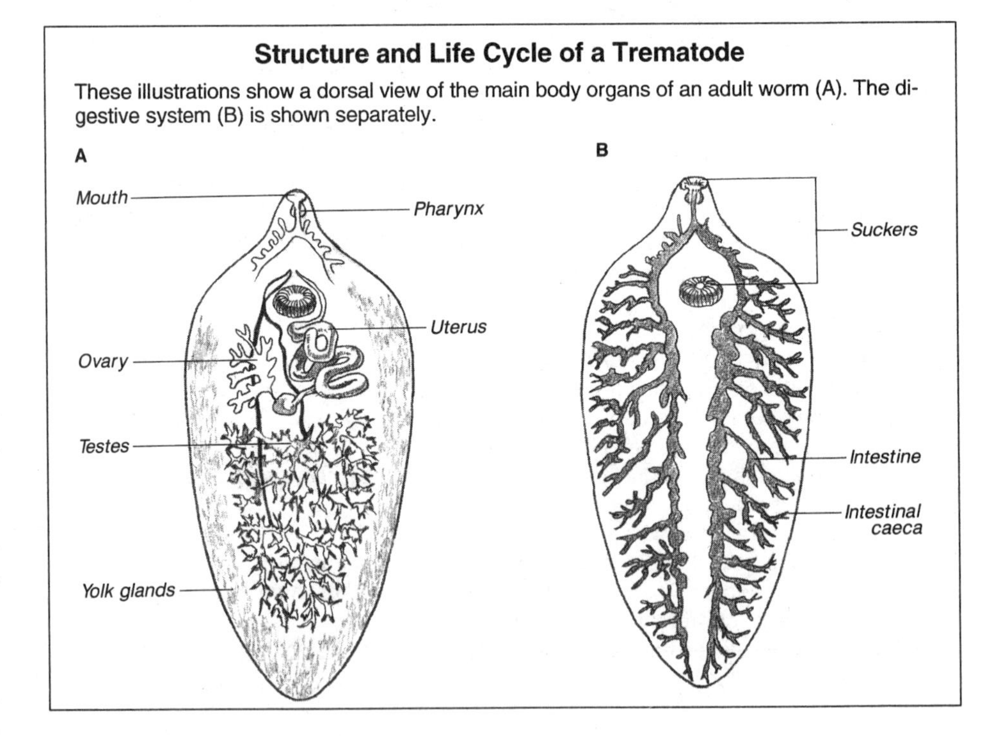

Structure and Life Cycle of a Trematode

These illustrations show a dorsal view of the main body organs of an adult worm (A). The digestive system (B) is shown separately.

7. The trematode life cycle is complex and has many stages
    a. Eggs are produced by adult worms in the definitive (final) host and are discharged in the host's feces, urine, or sputum
    b. The eggs remain dormant until they enter a body of water
    c. In the water, the eggs may be eaten by the intermediate host or they may hatch into ciliated *miracidium* larvae, which penetrate the intermediate host
    d. Several asexual generations occur in the intermediate host
    e. The last asexual generation produces a free-swimming larva called a *cercaria* or *metacercaria;* the larva enters the definitive host in a similar fashion to the miracidium larva
    f. In the definitive host, the larva matures into an adult worm
8. Trematodes cause debilitating and potentially fatal diseases in humans; they infect the circulatory system, lungs, bile ducts, pancreatic ducts, and intestines
9. Examples of flukes are *Schistosoma japonicum, S. mansoni,* and *S. haematobium* (which causes a disease called *schistosomiasis* that afflicts more than 300 million people worldwide); liver flukes, such as *Fasciola hepatica,* cause great economic loss in domesticated livestock

E. Class Cestoda

1. Cestodes, commonly called *tapeworms,* exist only as endoparasites in the gut of vertebrates; about 3,400 living species of tapeworms have been described
2. The class is characterized by a long, flat body composed of many reproductive segments, called *proglottids*
    a. The anterior end, called the *scolex,* is the organ of attachment; it is equipped with suckers, hooks, or both
    b. The *strobila,* or main body region, consists of a series of proglottids
        (1) New proglottids are produced just behind the scolex
        (2) As the segments mature, they are progressively pushed back toward the posterior end
        (3) Each proglottid contains a complete reproductive system with both male and female gonads
        (4) Egg-laden proglottids detach from the posterior end
        (5) Tapeworms can contain thousands of proglottids and reach 25 meters in length
3. Tapeworms do not have a digestive system; predigested nutrients from the host gut are absorbed through cellular projections covering the tegument
4. Sexual reproduction occurs by cross-fertilization or self-fertilization, depending on the availability of potential mates
5. Cestodes have complex life cycles involving one or more intermediate hosts
    a. Egg-laden proglottids detach from the posterior end of the strobila and are eliminated in the host feces
    b. In the pork tapeworm, the eggs are eaten by the intermediate host (a pig) and hatch into an *oncosphere* larva
    c. The oncosphere penetrates the host gut, migrates through the body, and encysts in muscle or connective tissue; the encysted stage is called a *cysticercus* or *bladder worm*
    d. If incompletely cooked pork is eaten by humans (the definitive host), the larva emerges into the gut and develops into an adult worm
    e. Humans also can serve as intermediate hosts for the pork tapeworm
        (1) The eggs may be ingested in contaminated food or water

(2) Juvenile tapeworms may encyst in the central nervous system, causing severe damage
(3) The cysts must be removed surgically

6. The beef *(Taeniarhynchus saginatus)* and pork *(Taenia solium)* tapeworms commonly infect humans; the dog tapeworm *(Echinococcus granulosus)* typically infests domestic dogs

VI. Basic Characteristics of Phylum Nemertea

A. General Information

1. Nemerteans, commonly called *ribbon worms,* are another acoelomate phylum
2. As with the flatworms, ribbon worms are triploblastic and bilaterally symmetrical; they range in size from microscopic to 60 meters in length
3. They have unsegmented, dorsoventrally flattened, elongated bodies that can stretch up to several times their body length; they may be brightly colored or patterned
4. About 900 living species of ribbon worms have been described

B. Ecologic relationships

1. Ribbon worms primarily are benthic marine animals, but some are planktonic
2. Microscopic individuals may be found in the interstitial spaces of subsurface marine sediments; a few species inhabit freshwater or moist terrestrial environments

VII. Nemertean Form and Function

A. General information

1. Nemerteans have a body plan similar to that of turbellarians
   a. The ciliated epidermis contains many gland cells
   b. Protonephridia with flame cells function in osmoregulation
   c. The nervous system and sense organs are like those of turbellarians
2. Ribbon worms also have several features that differ from those of flatworms
   a. They have a complete digestive system, with both mouth and anus
      (1) This is a significant evolutionary advance over animals with a gastrovascular cavity
      (2) Because food does not have to be regurgitated through the mouth, feeding and excretion can occur simultaneously, providing more frequent feeding opportunities
   b. They have a closed circulatory system; ribbon worms are the most primitive phylum with a circulatory system
   c. They are characterized by a special body cavity called a *rhynchocoel,* which contains a protrusible *proboscis;* the rhynchocoel and specialized proboscis are not found in any other animal phylum (see *Ribbon Worm*)

B. Locomotion

1. Small ribbon worms move by ciliary action, gliding over a slime trail laid down by gland cells in the epidermis

Ribbon Worm

A ribbon worm with proboscis extended to catch prey is depicted below.

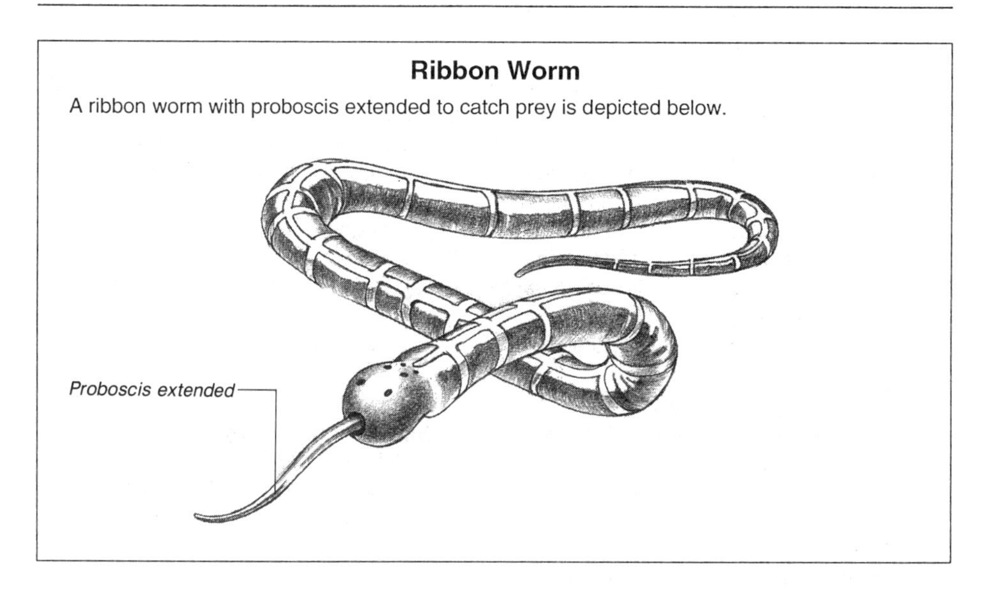

2. Larger worms move by contractions of body wall musculature; rhythmic body movements propel them through water or soft sediments

C. Nutrition

1. Most ribbon worms prey on other small invertebrates; some are scavengers and others may be herbivorous (details of feeding behavior are not known for many species)
   a. In predatory ribbon worms, a spike-tipped proboscis shoots out, harpoons the prey, and retracts into the mouth
   b. In species with no spike, the proboscis is covered with a sticky adhesive substance; the proboscis wraps around the prey and carries it to the mouth
2. Food is moved through the digestive tract by cilia and the action of body wall musculature
3. Digestion is similar to that seen in previous phyla, with both an extracellular and intracellular phase
4. Nutrients are distributed around the body by the circulatory system

D. Reproduction

1. Ribbon worms exhibit both sexual and asexual reproduction
2. Asexual reproduction occurs by multiple fission; each fragment develops into a new worm
3. Most ribbon worms are dioecious, but some are hermaphroditic
   a. In aquatic species, fertilization is external (gametes are released into the water); internal fertilization may occur in terrestrial species
   b. Eggs are released individually or embedded in a gelatinous or leathery egg case
   c. Aquatic forms usually have a free-swimming larval stage; the type of larva depends on the species

E. Circulation and gas exchange

1. Nemerteans have a closed circulatory system (all blood is contained within blood vessels and thin-walled spaces called *lacunae*)
2. Circulation is maintained by contraction of muscles in the blood vessel walls
3. The blood plasma contains pigmented cells that function in oxygen transport; some of these cells contain hemoglobin
4. White blood cells are present, but their function is unknown
5. Gas exchange occurs by diffusion across the body wall
   a. As with flatworms, the body form of ribbon worms facilitates diffusion to all body tissues
   b. Gases also are transported by the circulatory system

F. Excretion and osmoregulation

1. In contrast to flatworms, ribbon worm protonephridia are closely associated with the circulatory system and appear to function in excretion of metabolic wastes
2. Evidence also suggests that, as with flatworms, protonephridia play an important role in osmoregulation, especially in freshwater and terrestrial forms

G. Nervous system and sense organs

1. Ribbon worms exhibit a higher degree of cephalization than flatworms
2. The nervous system is composed of complex cerebral ganglia; impulses are carried by longitudinal nerve cords with ladder-like cross-connections
3. Ribbon worms possess a variety of sensory receptors, similar to those found in flatworms

Study Activities

1. List three basic characteristics of the acoelomate phyla.
2. Describe how flatworms carry out nutrition, gas exchange, and excretion.
3. Describe sexual and asexual reproduction in flatworms.
4. Create a chart that compares and contrasts the basic characteristics of each of the four flatworm classes.
5. Compare and contrast the basic characteristics of flatworms and ribbon worms.

8

Pseudocoelomate Phyla: Nematoda, Rotifera, and Others

Objectives

After studying this chapter, the reader should be able to:
- Describe the basic characteristics of the pseudocoelomate phyla.
- Explain current hypotheses concerning the evolutionary ancestry of the pseudocoelomates.
- Describe how nematodes perform their basic life functions.
- Compare and contrast the basic characteristics of nematodes with those of the rotifers and other pseudocoelomate phyla.

I. Basic Characteristics

A. General information

1. The *pseudocoel* (a fluid-filled body cavity between the digestive tract and body wall) is derived from the embryonic blastocoel; it is not lined with peritoneum (see Chapter 3, Body Structure and Function)
2. Representative pseudocoelomate phyla are Nematoda, Rotifera, Gastrotricha, Kinorhyncha, and Loricifera

B. Ecologic relationships

1. The pseudocoelomates represent several diverse phyla and thus have very different life-styles
2. This group includes both free-living and parasitic species
3. Although small, pseudocoelomates are numerous and ecologically important inhabitants of all aquatic and terrestrial habitats

C. Evolutionary relationships

1. Evolutionary relationships among the pseudocoelomate phyla are difficult to determine, because some phyla contain no free-living individuals
2. Current hypotheses suggest that pseudocoelomates are descended from the protostome evolutionary line (see Chapter 4, Classification and Phylogeny of Animals)
3. The mode of development, small size, ciliation patterns, and presence of protonephridia in adults lead many experts to suggest that pseudocoelomates arose by neoteny from coelomate ancestors

4. At one time, all pseudocoelomates were grouped together into one phylum, Aschelminthes; differences in morphology and physiology have led to the separation of this group into individual phyla

II. Basic Characteristics of Phylum Nematoda

A. General information

1. Nematodes are elongate, unsegmented worms with bilateral symmetry; they are commonly called *roundworms* or *threadworms*
2. These triploblastic animals have a body that is round in cross section and covered with a resistant, nonliving cuticle
3. Muscles in the body wall run longitudinally only; there are no circular muscle layers

B. Ecologic relationships

1. Nematodes are among the most abundant metazoans
   a. One square meter of farming soil may contain 3 million nematodes
   b. More than 90,000 individuals have been recovered from a single rotting apple
2. Over 12,000 living species have been identified, but experts estimate several times that number remain to be described (estimates run as high as 500,000 potential species)
3. Many nematodes are parasitic, causing serious human diseases and economic problems with infestations of crops and domestic animals
4. Free-living forms are common inhabitants of interstitial spaces in both terrestrial and aquatic sediments

III. Nematode Form and Function

A. General information

1. Roundworms are, as the name implies, cylindrical in cross section
2. Most are small (microscopic to 5 cm in length), but some species exceed a meter in length
3. The pseudocoel functions as a hydrostatic skeleton (see Chapter 3, Body Structure and Function)

B. Body types of roundworms

1. Nematodes usually are small, and they exhibit bilateral symmetry
2. The elongate, slender, cylindrical body frequently is tapered at the anterior and posterior ends (see *Representative Pseudocoelomate Phyla*)
3. The body is covered by a thick, noncellular, collagenous ***cuticle*** secreted by the epidermis
   a. The cuticle allows nematodes to resist environmental stress (such as dry terrestrial habitats or, in endoparasites, host digestive fluids)
   b. It is molted for growth

Representative Pseudocoelomate Phyla

The following diagrams show the structural differences between a nematode, a rotifer, a gastrotrich, a kinorhynch, and a loriciferan.

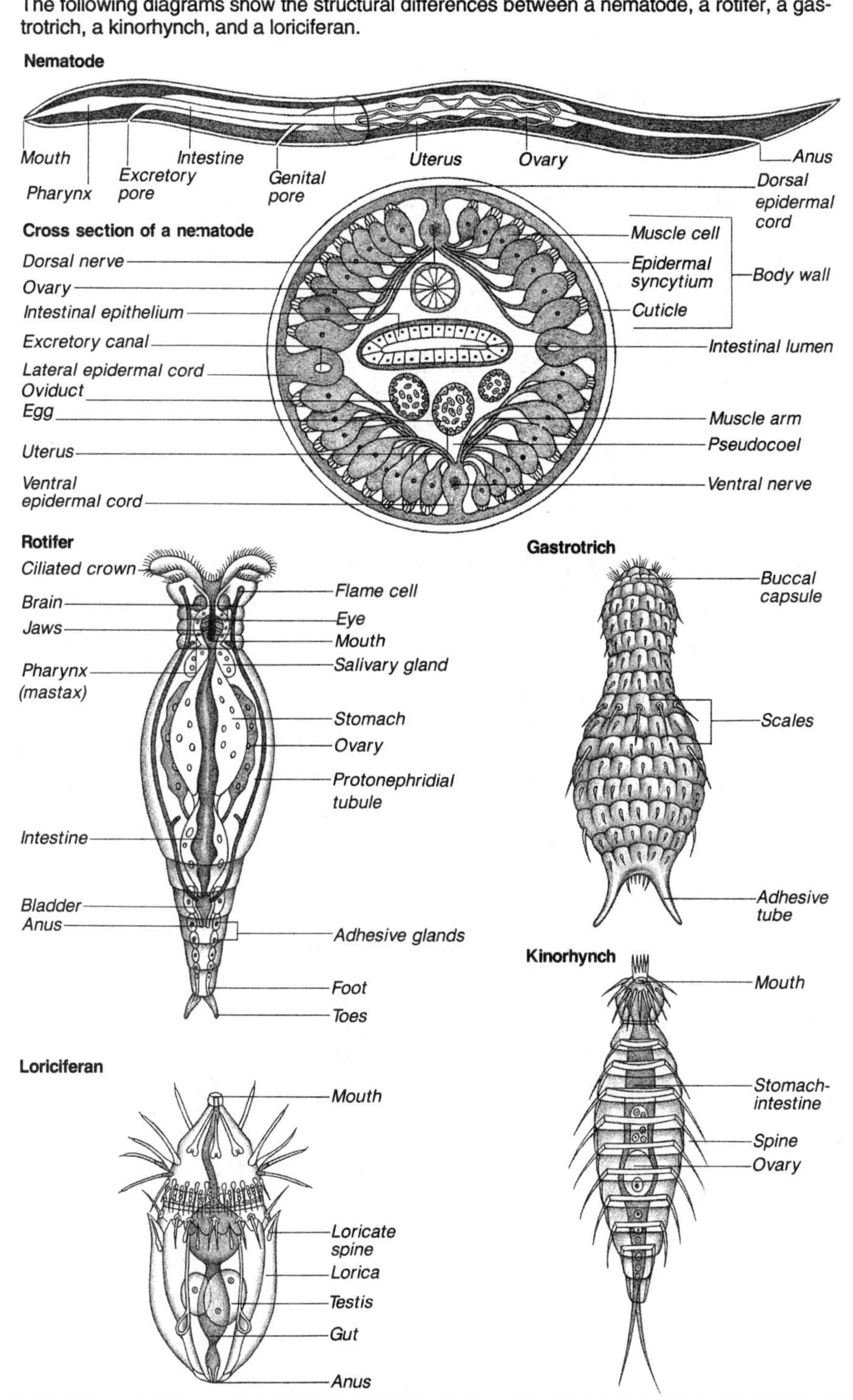

IV. Internal Structure and Physiology of Roundworms

A. General information

1. Nematodes have only longitudinal muscle layers in the body wall, arranged in four bands
2. Fluid in the small pseudocoel acts as a hydrostatic skeleton and supports the internal organs

B. Locomotion

1. Roundworms do not have cilia or flagella; they move by contracting the body wall musculature
2. Nematode locomotion is initiated by contraction of the longitudinal muscles, which push against the cuticle
   a. Forward movement is accomplished by pushing the body against the substrate
   b. Lacking a suitable substrate, forward motion is difficult; in an aquatic environment, nematodes move with a characteristic side-to-side thrashing motion
3. The longitudinal muscle fibers are connected to the dorsal and ventral nerve cords by long extensions called *muscle arms* (see *Representative Pseudocoelomate Phyla,* page 67)

C. Nutrition

1. Free-living nematodes are carnivorous predators or scavengers with a broad diet; parasitic nematodes feed on host tissues or body fluids
   a. Evidence suggests that many scavengers do not feed directly on dead bodies, but on the bacteria and fungi in the decomposing organic matter
   b. Their prey include most invertebrates that are small enough to be captured and consumed; many roundworm species are herbivorous
2. Food is sucked through the mouth into the pharynx
3. Nematodes have a complete digestive tract, which varies in length and complexity; digestive tract features are important for taxonomic classification within this group
4. Initial digestion is extracellular, and absorption of nutrients occurs in the midgut region

D. Reproduction

1. Roundworms reproduce sexually, and fertilization is internal
2. Most nematodes are dioecious and show *sexual dimorphism* (the male is smaller than the female and is distinctly curved at the posterior end)
3. The male reproductive system consists of one or more testes
   a. The testes are drained by sperm ducts, which lead into a seminal vesicle; sperm are produced and stored in the seminal vesicle before copulation
   b. The seminal vesicle connects with an ejaculatory duct; this duct connects with the anus, which functions in both reproduction and excretion
   c. In many species, seminal fluids from prostatic glands empty into the ejaculatory duct
   d. Most male nematodes have two copulatory spicules at the posterior end; these are inserted into the female *gonopore* (copulatory opening) for sperm transfer

4. The female reproductive system consists of paired ovaries (see *Representative Pseudocoelomate Phyla,* page 67)
   a. The ovaries narrow into a pair of oviducts, which then enlarge into paired uteri
   b. The two uteri join together to form a vagina that opens into the gonopore on the midventral surface; the gonopore is separate from the anus in females
   c. The female lays double-shelled eggs; free-living forms deposit their eggs in the sediment
5. Free-living nematodes usually have direct development; there are four juvenile stages, each of which is separated by molting of the cuticle
6. Parasitic forms have complex life cycles involving one or more intermediate hosts

E. Circulation, gas exchange, excretion, and osmoregulation

1. Parasitic nematodes frequently are facultative (optional) or obligate (required) anaerobes; free-living forms and the free-living larval stages of parasitic species usually are obligate aerobes
2. There are no special organs for circulation or gas exchange; these functions take place by diffusion and movement of pseudocoelomic fluids
3. Nematodes have a unique osmoregulatory system of glands and collecting tubules; this system functions to a lesser degree for excretion
4. The primary excretory product is *ammonia;* ***urea*** also is produced in hypertonic environments

F. Nervous system and sense organs

1. The cerebral ganglia form a ring around the esophagus
2. Several longitudinal nerve cords originate in the esophageal nerve ring and run the length of the body
   a. The major nerve cord is ventral and includes both motor and sensory nerves
   b. The dorsal nerve cord contains only motor nerves
   c. The lateral nerves primarily are sensory
3. Nematodes have numerous sensory receptors distributed along the body
   a. Tactile information is critical to nematodes, because most species inhabit interstitial spaces or are endoparasites
   b. Chemosensory receptors also are abundant

V. Groups of Parasitic Nematodes

A. General information

1. Nematodes parasitize many vertebrate and invertebrate animals
2. Several groups of parasitic nematodes are medically and economically important
3. The groups are separated by the details of their life cycles, mode of transmission, and symptoms of infection

B. Ascaroids

1. Ascaroids are endoparasites of many vertebrates, including humans; nearly 1.26 billion humans are infected worldwide
2. The life cycle is simple and confined to one host
   a. Adults live in the host's small intestine
   b. Eggs are passed in the feces

(1) The eggs are extremely resistant to adverse environmental conditions
(2) They can survive for many years in the soil
c. Infection occurs through ingestion of the eggs
(1) They may be eaten with uncooked vegetables
(2) Children often are infected by placing their hands or other contaminated objects in their mouths
d. Juvenile ascaroids hatch in the intestine
(1) They penetrate the intestinal wall and enter the circulatory system, where they are carried by the blood to the lungs
(2) They migrate through the alveoli and return to the intestine through the bronchi, trachea, and esophagus; they may seriously damage the lung tissue or cause pneumonia during this stage of the life cycle
e. Adult worms feed on intestinal contents; in large numbers, they can block the intestines
3. *Ascaris lumbricoides* is one of the best-known parasitic nematodes that infect humans
a. Infection occurs worldwide and is common in the southeastern United States
b. *Ascaris* is among the larger nematodes, reaching a length of 49 centimeters

C. Hookworms
1. Hookworms are characterized by a dorsal curve (hook) in the anterior end
2. They live in the host's intestines and feed on blood
a. Sharp mouth plates cut through the intestinal lining, causing hemorrhage
b. Hookworms consume large quantities of blood, and massive infections can cause serious anemia in the host
3. Eggs are passed in the feces
a. The eggs hatch in the soil, where the juveniles feed on bacteria
b. Juvenile hookworms penetrate human skin and enter the bloodstream
c. They reach the lungs and migrate to the intestine in a manner similar to that of ascaroids
4. *Necator americanus* is a typical example

D. Filaroids
1. Filarial worms are endoparasites of vertebrates, especially birds and mammals; more than 250 million humans worldwide are infected by these parasites
2. The worms live in the lymphatic system
3. Heavy infections can block lymph vessels, causing severe edema and gross enlargement of the extremities (primarily the arms, legs, and scrotum); this disease is known as *elephantiasis*
4. The life cycle requires an arthropod intermediate host; bloodsucking insects, such as mosquitos, fleas, and biting flies, are common ***vectors*** (transmitters)
a. Female worms produce live young, called *microfilariae,* which are ingested by bloodsucking insects as they feed
b. Humans are infected through the bite of an infected insect
5. Common filaroids are *Wuchereria bancrofti,* which causes elephantiasis, and *Onchocerca volvulus,* which causes river blindness

E. Trichinellids
1. Trichina worms infect humans and other mammals, causing a serious disease called *trichinosis*

2. Adult worms live in the small intestine; females burrow into the intestinal lining and give birth to live young
   a. Juveniles penetrate into blood vessels and migrate to skeletal muscles throughout the body
   b. The worms encyst in the skeletal muscle
   c. The cysts must be eaten by another mammal host to continue the infective cycle
   d. When meat that contains cysts with live worms is eaten, the juveniles emerge into the gut and develop into adult worms
3. As with tapeworm infection, humans usually are infected by eating incompletely cooked pork; pigs become infected by eating garbage with cyst-containing meat scraps or by eating infected rats

F. Pinworms

1. Pinworms do not cause life-threatening illness, but they are the most common worm parasites in the United States; authorities estimate that 30% of children and 16% of adults in the U.S. are infected with pinworms
2. Adult worms live in the large intestine
   a. Females travel to the anal region at night to lay their eggs
   b. Infection results from ingestion of the eggs
      (1) Worm activity in the anal region causes intense itching; scratching contaminates the hands
      (2) Adult humans may inhale the eggs when shaking out sheets and blankets
   c. Eggs are swallowed, then hatched in the duodenum (the initial segment of the small intestine)
   d. The worms migrate to the large intestine and develop into adults

VI. Basic Characteristics of Phylum Rotifera

A. General information

1. Members of phylum Rotifera commonly are called *rotifers*
2. More than 1,800 living species have been described
3. Most are quite small, ranging from 1 to 3 millimeters in length
4. The phylum is characterized by a rotating crown of cilia, from which the group takes its name
5. Two adhesive toes, used for attachment to the substrate, are found at the posterior end

B. Ecologic relationships

1. Rotifers live in freshwater habitats, including the moist film on soil or vegetation; a few species are marine
2. They may be planktonic, sessile, or actively motile; most are benthic, and some live in the interstitial spaces of sediments
3. In adverse environmental conditions, rotifers may enter a resistant, dormant state called ***cryptobiosis***

VII. Rotifer Form and Function

A. General information

1. Rotifers are triploblastic and unsegmented, with bilateral symmetry
2. The digestive tract is complete and contains specialized regions
3. The body has three sections: head, trunk, and foot (see *Representative Pseudocoelomate Phyla,* page 67)
4. The rotating crown of cilia at the anterior end, called the *corona*, functions in feeding
5. The body surface is covered with a protective cuticle, and the body wall contains circular and longitudinal muscle layers
6. The internal organs lie within a large, fluid-filled pseudocoel
    a. The pseudocoel also functions as a hydrostatic skeleton
    b. Changes in hydrostatic pressure can be used to extend body parts

B. Locomotion

1. Motile rotifers move by swimming or creeping
2. Swimming is accomplished by the action of the coronal cilia
3. Creeping is a stepwise process
    a. The rotifer attaches its posterior end, or foot, to the substrate with adhesive glands
    b. The anterior end elongates and attaches to the substrate in the direction of travel
    c. The foot is released and dragged forward

C. Nutrition

1. Rotifers are predatory or filter feeders
    a. The beating of coronal cilia draws water into the mouth
    b. The mouth opens into a muscular pharynx, called a *mastax,* that is unique to this phylum
        (1) The mastax has jaws with teeth that grind up the prey
        (2) The teeth vary in shape and size, depending on the diet
        (3) In predatory species, the mastax may be modified to pierce the body wall of plants or animals; the rotifer then feeds on the prey's body fluids
2. Rotifers have a complete digestive tract
    a. Digestion is extracellular
    b. Digestion and absorption occur in the stomach

D. Reproduction

1. Rotifers reproduce sexually
2. These dioecious animals exhibit sexual dimorphism
    a. Males are smaller and less complex than females
    b. Males constitute only a small percentage of the population
3. In species with males, fertilization is internal; the male copulatory organ is inserted into the female cloaca or sperm are injected hypodermically through the body wall
4. Some species have only female members and reproduce by ***parthenogenesis;*** these females produce diploid, unfertilized eggs that develop into other females only

5. Most rotifers can produce two different types of eggs, *amictic* (diploid eggs) and *mictic* (haploid eggs that have undergone meiosis)
   a. If mictic eggs are not fertilized, they rapidly develop into males
   b. If these eggs are fertilized, they develop a thick shell and become dormant for several months; at the end of that period, they activate and develop into females
   c. Mictic eggs are resistant forms that are produced to withstand adverse environmental conditions; they commonly are produced in freshwater habitats to survive cold or dry seasons

E. Circulation, gas exchange, excretion, and osmoregulation
1. Rotifers have no special organs for circulation or gas exchange
   a. Gas exchange occurs by diffusion across the body wall
   b. Gases and nutrients are distributed in the pseudocoelomic fluid
2. Protonephridia with flame cells function in osmoregulation and excretion of metabolic wastes

F. Nervous system and sense organs
1. The cerebral ganglia are located anterior to the mastax
   a. Several longitudinal nerve tracts arise from these ganglia
   b. Both main tracts may lie ventrally, or one may lie dorsally and one ventrally, depending on the species
2. Sensory receptors of several types are distributed over the body surface

VIII. Basic Characteristics of Other Pseudocoelomate Phyla

A. General information
1. Although small in size and number of species, representatives of the other pseudocoelomate phyla have great ecologic significance
2. They exist in large numbers and are important inhabitants of interstitial communities

B. Phylum Gastrotricha
1. About 400 living species of gastrotrichs have been described
2. Members of this phylum exhibit several adaptations for protection in an interstitial life-style (see *Representative Pseudocoelomate Phyla,* page 67)
   a. The cavity of the pseudocoel is filled with mesenchyme tissue, rendering gastrotrichs effectively acoelomate
   b. As with many interstitial animals, gastrotrichs have many adhesive tubes for attachment to the substrate
   c. The cuticle is well developed; plates and spines protect the animal from the crushing action of sand grains
   d. Most sensory receptors are tactile receptors or chemosensors
3. Gastrotrichs are unsegmented, with bilateral symmetry
4. Most gastrotrich species are hermaphroditic; the remainder have only parthenogenetic females
5. They have no special organs for circulation or gas exchange
6. Protonephridia are found only in freshwater species, suggesting that their primary function is osmoregulation

C. Phylum Kinorhyncha

1. Kinorhyncha is a small phylum (about 100 species)
2. Most kinorhynchs are interstitial inhabitants of marine sediments
3. These animals are segmented internally and externally; the body is divided into 13 segments (see *Representative Pseudocoelomate Phyla,* page 67)
4. Kinorhynchs are bilaterally symmetrical
5. They have structural support features similar to those of gastrotrichs
    a. The cuticle is well developed with rigid, protective plates
    b. Spines function in protection and locomotion
6. Little is known about feeding and digestion in kinorhynchs; they are presumed to extract organic material and unicellular algae from the sediments
7. There are no special organs for circulation or gas exchange; the details of excretory and osmoregulatory functions are unknown
8. The nervous system is simple, with sensory receptors distributed over the body surface
9. Kinorhynchs reproduce sexually; males and females are similar in appearance, but the details of reproduction and development are unknown

D. Phylum Loricifera

1. Loriciferans are the most recently discovered phylum in the animal kingdom (discovered in 1983)
2. Except for one brief glimpse of a loriciferan larva, no member of this phylum has ever been seen alive
3. The phylum is named for a protective cuticle composed of six plates, called a *lorica*
4. Like the gastrotrichs and kinorhynchs, loriciferans display protective adaptations for an interstitial habitat (see *Representative Pseudocoelomate Phyla,* page 67)
    a. The anterior end of the body is retractable into the abdomen, which is protected by the lorica
    b. Most species have a pair of adhesive toes
    c. Spines and protective plates are distributed over the body
5. The body is unsegmented with bilateral symmetry; it is divided into a head, thorax, and abdomen
6. No special organs for circulation or gas exchange have been identified
7. Protonephridia are present, but the details of their function are unknown
8. Loriciferans have separate sexes; methods of reproduction and development are unknown, but at least one larval form has been identified

Study Activities

1. List three basic characteristics of the pseudocoelomate phyla.
2. Describe how nematodes and rotifers obtain nutrition, perform gas exchange, and excrete wastes.
3. Describe methods of reproduction in nematodes and rotifers.
4. Create a chart that compares and contrasts the basic characteristics of nematodes with those of other pseudocoelomate phyla.
5. Describe three structural adaptations for interstitial habitats.

9

Phylum Mollusca

Objectives

After studying this chapter, the reader should be able to:
- Describe the basic characteristics of molluscs.
- Explain current hypotheses concerning the evolutionary ancestry of molluscs.
- Describe how molluscs perform their basic life functions.
- Identify and characterize the classes of phylum Mollusca.

I. Basic Characteristics

A. General information

1. Molluscs are bilaterally symmetrical protostomes and have a true coelom
   a. The coelom is reduced in size
   b. The main body cavity is part of an open circulatory space, called a *hemocoel*
2. Molluscs have soft bodies that usually are protected by a secreted shell
3. Most molluscs move by means of a ventral muscular foot
4. Members of the eight classes of molluscs differ in external appearance but share the same general body plan

B. Ecologic relationships

1. Molluscs are found in all aquatic and several terrestrial habitats
2. Most molluscs are marine, with a wide variety of life-styles; both active and sedentary species have been identified
3. With more than 100,000 living species, phylum Mollusca is one of the largest animal phyla; about 60,000 additional species have been described from the fossil record
4. Symbiotic relationships between molluscs and other animals are common
   a. Representatives of several phyla attach to gastropod and bivalve mollusc shells
      (1) Frequently seen symbiotes include sponges, tube worms, ectoprocts, hydroids, and anemones
      (2) In this mutualistic relationship, the mollusc receives camouflage from predators and, with cnidarian symbiotes, protection from the cnidarian's stinging tentacles; the "hitchhikers" receive transportation and greater availability of food resources
   b. As with anthozoan cnidarians, the giant clam, *Tridacna,* has a close mutualistic relationship with unicellular algae (zooxanthellae) living inside the

clam's body tissues; the algae receive protection and exposure to sunlight and access to the nitrogenous waste products of the host, and the clam receives organic compounds produced by photosynthesis

c. A recently discovered form of symbiosis exists between molluscs of deep-sea geothermal habitats and sulfur bacteria; details of this relationship are unclear; however, the clams appear to derive a significant portion of their nutritional needs from the bacteria

5. Molluscs are of great economic importance
   a. Molluscs from several classes are principal sources of food
   b. The worldwide squid and octopus fisheries alone produce more than 2 million metric tons annually
   c. A huge jewelry industry is based on natural and cultured pearls produced by bivalve molluscs

C. Evolutionary relationships

1. Until recently, molluscan phylogeny was based on a hypothetical ancestral mollusc
   a. The molluscan phylogenetic tree was constructed by determining the degree of divergence from this hypothetical ancestor
   b. Descent from the hypothetical ancestor cannot account for all living molluscan groups; therefore, this approach no longer is considered valid
2. At present, contradictory theories of molluscan evolution have not been resolved
   a. The *turbellarian theory* states that molluscs arose from a turbellarian flatworm, which also gave rise to the other main protostome line, annelids and arthropods
      (1) Molluscs and turbellarians have similar locomotion patterns; both use ventral ciliary gliding
      (2) Molluscs have spiral ***cleavage,*** trochophore larva, and other protostome characteristics
   b. The *coelomate theory* proposes that molluscs arose with the annelids from a coelomate, segmented ancestor
      (1) Molluscs have spiral cleavage, trochophore larvae, and other protostome characteristics
      (2) Primitive molluscs have large pericardial (heart) cavities, which suggests descent from a coelomate ancestor
   c. The *annelid theory* suggests that molluscs evolved from the annelid line with a secondary loss of segmentation and other annelid characters; this theory is not widely accepted

II. Mollusc Form and Function

A. General information

1. Molluscs are bilaterally symmetrical or secondarily asymmetrical
2. These unsegmented animals usually have a well-defined head
3. The ventral body wall is specialized into a muscular foot, and most species have a protective shell
4. The dorsal body wall is specialized into a *mantle,* which covers the internal organs and secretes the shell

5. The coelom is reduced in size and limited to a small area around the heart (the *pericardial cavity*)
6. The circulatory system is open (see Chapter 3, Body Structure and Function); interconnected blood sinuses (the hemocoel) form the main body cavity
7. One to many pairs of *metanephridia* (also called kidneys) function in excretion and osmoregulation

B. Body types of molluscs

1. All molluscs exhibit variations on a generalized body plan consisting of three main regions: the head, the visceral mass, and the mantle
2. The primary body cavity is the hemocoel, an open circulatory space composed of interconnected blood sinuses
3. Most species are protected by a single or bivalved calcium carbonate shell secreted by the mantle
4. The gut is complete with specialized regions

III. Internal Structure and Physiology of Molluscs

A. General information

1. The body wall of molluscs has three layers: the cuticle, the epidermis, and the muscles
2. The dorsal body wall is modified into the mantle; this living tissue grows with the individual mollusc
   a. A cavity formed by the folds in the mantle houses and protects the internal organs
   b. Beating of cilia creates a water current through the mantle cavity, carrying gametes and metabolic wastes out and allowing food and oxygen to enter
   c. In cephalopods, the shell is reduced or absent
      (1) The exposed fleshy body surface is the mantle
      (2) The mantle is modified into a tubular funnel, called a *siphon*
      (3) Cephalopods move by water-powered jet propulsion; water from the mantle cavity is forcibly expelled through the siphon
   d. Many molluscs can retract their bodies into the mantle cavity, which is protected by a shell
3. Mollusc shells vary in shape and size, but all are constructed of layers of calcium carbonate
   a. The shell increases in size as the animal grows; new layers are added at the inner or outer shell margin
   b. Calcium for shell construction is extracted from the surrounding water or soil, or from food
4. Gastropods undergo a unique twisting growth pattern, called *torsion*
   a. Torsion involves a counterclockwise rotation of the entire body (up to 180 degrees), including the visceral mass, mantle, and shell
   b. This two-step process occurs during development
   c. The spiral shell of gastropods is *not* the result of torsion; this occurs during early embryonic development, before torsion occurs
   d. Torsion allows the head end of the animal to be drawn into the mantle cavity for protection

e. On the negative side, the anus and excretory pore in torted animals are positioned to drop wastes on the head and gills; this is referred to as *fouling*
f. Gastropods have structural adaptations to avoid fouling
5. The muscular foot of molluscs may be adapted for locomotion or attachment to the substrate
a. The foot is extended and anchored by engorgement with *hemolymph* (circulatory fluid in the hemocoel); the body is pulled forward by muscular contraction
b. The extended foot also is used for boring and burrowing
c. In permanently attached molluscs, the foot may be reduced or absent
d. In some species, such as nudibranchs (nonshelled gastropods), the foot is modified into fins for swimming
e. In other species, such as the violet snail *(Janthina)*, the foot secretes a raft of bubbles for flotation

B. Nutrition
1. Molluscs are herbivores, predatory carnivores, or filter feeders
2. The *radula,* a tongue-like organ with rasping, chitinous teeth, is found in all molluscs except bivalves
3. In herbivores, the radula is used for rasping or tearing
a. Intertidal molluscs use the radula to scrape algae from rock surfaces
b. Abalone hold seaweed with the foot, tearing off pieces with the radula
4. In predatory carnivores, the radula may be adapted for piercing, tearing, or cutting
a. Some species, such as predatory whelks (gastropods) and octopuses (cephalopods), use the radula to bore holes in the shells of bivalve prey; this boring action may be supplemented by secretion of acidic chemicals that dissolve the calcium carbonate of the prey's shell
b. In carnivores that swallow their prey whole, the radula is modified into long, harpoon-like teeth, which seize the prey and pull it to the mouth
c. In cone snails (gastropods), the radula is modified into a hypodermic structure that injects a paralyzing venom into the prey
d. Cephalopods have strong, beak-like jaws that bite and hold the prey, which is pulled into the mouth with the radula; many cephalopods secrete a paralyzing venom from their salivary glands
5. Filter feeders, such as bivalves, secrete mucus to trap food particles carried by water currents into the gill cavity
a. In most species, the gill filaments, called *ctenidia*, function in both gas exchange and feeding
b. Most filter feeders have a reduced or absent radula
6. Molluscs have complete digestive tracts
a. Ingested food is carried through the esophagus to the stomach
b. Extracellular digestion takes place in the stomach, *ceca* (pouch-like extensions of the stomach wall), or digestive glands, facilitated by digestive enzymes secreted by the stomach and accessory glands
c. Filter-feeding molluscs may have a stomach adapted for sorting fine particles
(1) Small particles are carried into the stomach and digestive ceca
(2) Rejected larger particles are carried along ciliated grooves to the intestine for excretion
d. Absorption occurs primarily in the intestine

C. Reproduction

1. Molluscs reproduce sexually; most are dioecious, but gastropods may be hermaphroditic
2. Fertilization can be either internal or external, depending on the species
3. Eggs are shed into the environment; they frequently are enclosed in gelatinous masses or strings
4. Development may be direct or indirect
   a. In indirect development, a ciliated, free-swimming *trochophore* larva (similar to that of annelids) is produced
   b. In some molluscs, the trochophore is the only larval stage
   c. In others, a second *veliger* larva is produced; the veliger may have a foot, a shell, an operculum, and other adult characteristics
      (1) Some freshwater bivalves produce parasitic veliger larvae called *glochidia;* these larvae are ectoparasites of fish, absorbing nutrients from the host tissues
      (2) Gastropod torsion usually occurs during the veliger larva stage
   d. Cephalopod development always is direct

D. Circulation and gas exchange

1. The main body cavity of molluscs is the hemocoel, an interconnected system of large venous sinuses
2. Molluscs have a dorsal, three-chambered heart; it has two *atria* (sometimes called *auricles*) and a single *ventricle*
3. The blood of molluscs is called *hemolymph*
   a. The main respiratory pigment, hemocyanin, is dissolved in the hemolymph
   b. Many molluscs also have hemoglobin and myoglobin
   c. The hemolymph contains various types of cells
4. Hemolymph circulates from the heart, through vessels, and into the sinuses of the open circulatory system (see *Circulatory System of a Bivalve Mollusc,* page 80)
   a. Oxygenated hemolymph travels from the ctenidia to the atria in efferent branchial vessels
   b. The muscular ventricle pumps the hemolymph into a large anterior artery, which branches and drains into the sinuses
   c. Within the sinuses, the tissues are bathed in oxygenated hemolymph
   d. Hemolymph from the sinuses returns to the ctenidia through afferent branchial vessels
5. Gas exchange takes place in the branchial blood vessels
   a. Most molluscs have ctenidia
   b. Those that do not rely primarily on gas exchange across the mantle or general body surface

E. Excretion and osmoregulation

1. The basic excretory structures of molluscs are paired, tubular metanephridia (often called kidneys)
   a. The kidneys are large sacs with numerous folds that increase the surface area
   b. Wastes are excreted through a nephridiopore in the mantle cavity
      (1) In many molluscs (especially gastropods), the nephridiopore also functions as a *urogenital pore*
      (2) Both excretory wastes and gametes are expelled from this opening

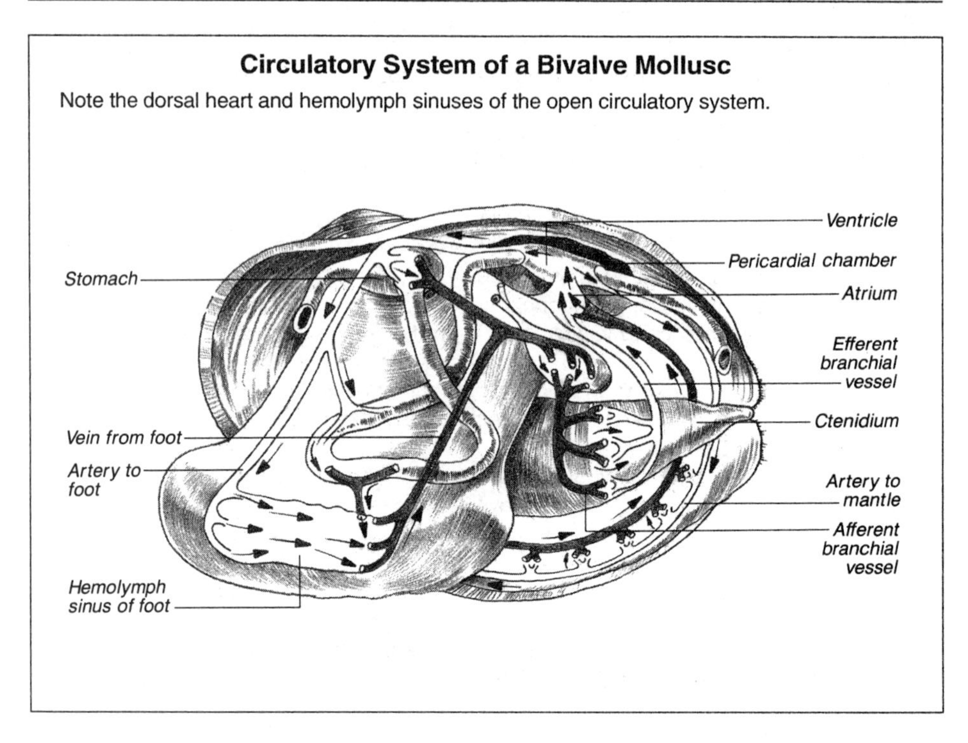

Circulatory System of a Bivalve Mollusc

Note the dorsal heart and hemolymph sinuses of the open circulatory system.

2. Molluscs may have one to several pairs of nephridia
   a. Aquatic molluscs primarily are osmoconformers and excrete ammonia
   b. Terrestrial species excrete water-conserving uric acid

F. Nervous system and sense organs

1. In most molluscs, the nervous system consists of three pairs of large ganglia, ringed around the digestive tract at the anterior end of the body
2. Two pairs of longitudinal nerve cords arise from these anterior ganglia; the nerve cords are located ventrally and have ladder-like cross-connections (similar to the flatworm pattern)
3. Cephalopods have the highest nervous system development of all invertebrates
   a. Most of the ganglia in cephalopods are located at the extreme anterior end of the animal
   b. The ganglia interconnect into lobes of a large brain
   c. A large optic nerve extends to each eye; cephalopods have excellent vision, comparable to that of vertebrates
   d. Cephalopods have a large behavioral repertoire and are excellent learners
      (1) They have well-developed rapid-escape behaviors
      (2) They have extensive mimicking and camouflage abilities
      (3) They can master complex memory-dependent tasks
4. Molluscs possess a wide variety of sensory receptors
   a. Sensory tentacles and photoreceptors are located at the anterior end
   b. Patches of chemosensory epithelium, called *osphradia*, are found on the gills
   c. Chemosensors also are located elsewhere on the body; some species (especially nudibranchs) use a form of chemical interspecific communication
   d. Statocysts and geomagnetic sensors may also be present

IV. Classification of Phylum Mollusca

A. General information

1. Molluscs are differentiated by shell type, morphology, and life-style
2. In many cases, however, taxonomic decisions have been made based on the shell alone

B. Class Caudofoveata

1. Members of this class are small, worm-like, nonshelled inhabitants of deep sea interstitial habitats
2. The body wall has a cuticle with embedded calcareous spicules; the foot, eyes, tentacles, and statocysts are absent
3. About 70 living species have been described; however, little is known about their biology or life-style
4. Examples include *Limifossor* and *Chaetoderma*

C. Class Solenogastres (Aplacophora)

1. These small (usually microscopic), elongated molluscs do not have shells (see *Representative Molluscs,* page 82)
    a. The body wall has a cuticle with embedded calcareous spicules
    b. The foot is absent, but a ventral furrow (called a *pedal groove*) is thought to function in locomotion
    c. Eyes, tentacles, statocysts, and metanephridia are absent; ctenidia and radula may be absent
2. About 250 living species have been described; many feed on cnidarians
3. They are found in deep sea (greater than 200 meters) interstitial or benthic habitats
4. Examples include *Kruppomenia* and *Proneomenia*

D. Class Monoplacophora

1. Until 1952, monoplacophorans were known only from the fossil record; since that time, 11 living species have been described
2. All are small (about 3 centimeters long) and live in deep sea habitats
    a. They have a single dorsal shell that is peaked at the anterior end
    b. The ventral surface is broad and flat
    c. The mantle cavity is reduced to a shallow groove around the foot
    d. Internal organs include five or six pairs of ctenidia, two pairs of gonads, and six or seven pairs of metanephridia
    e. The head is small but distinct
    f. Eyes are absent; tentacles are located around the mouth
3. Little is known about the biology or life-style of this class of molluscs
4. Examples include *Neopilina, Monoplacophorus,* and *Vema*

E. Class Polyplacophora

1. Polyplacophorans (commonly called *chitons*) are flattened, elongated molluscs with a broad ventral foot (see *Representative Molluscs,* page 82)
    a. The class is characterized by a dorsal shell composed of eight overlapping plates
    b. The mantle forms a girdle around the shell margin; in some species, the mantle covers part or all of the shell plates

Representative Molluscs

The illustrations below show the major body structures for five types of molluscs: solenogastre, chiton, tusk shell, clam, and squid.

Solenogastre

Chiton

Tusk shell

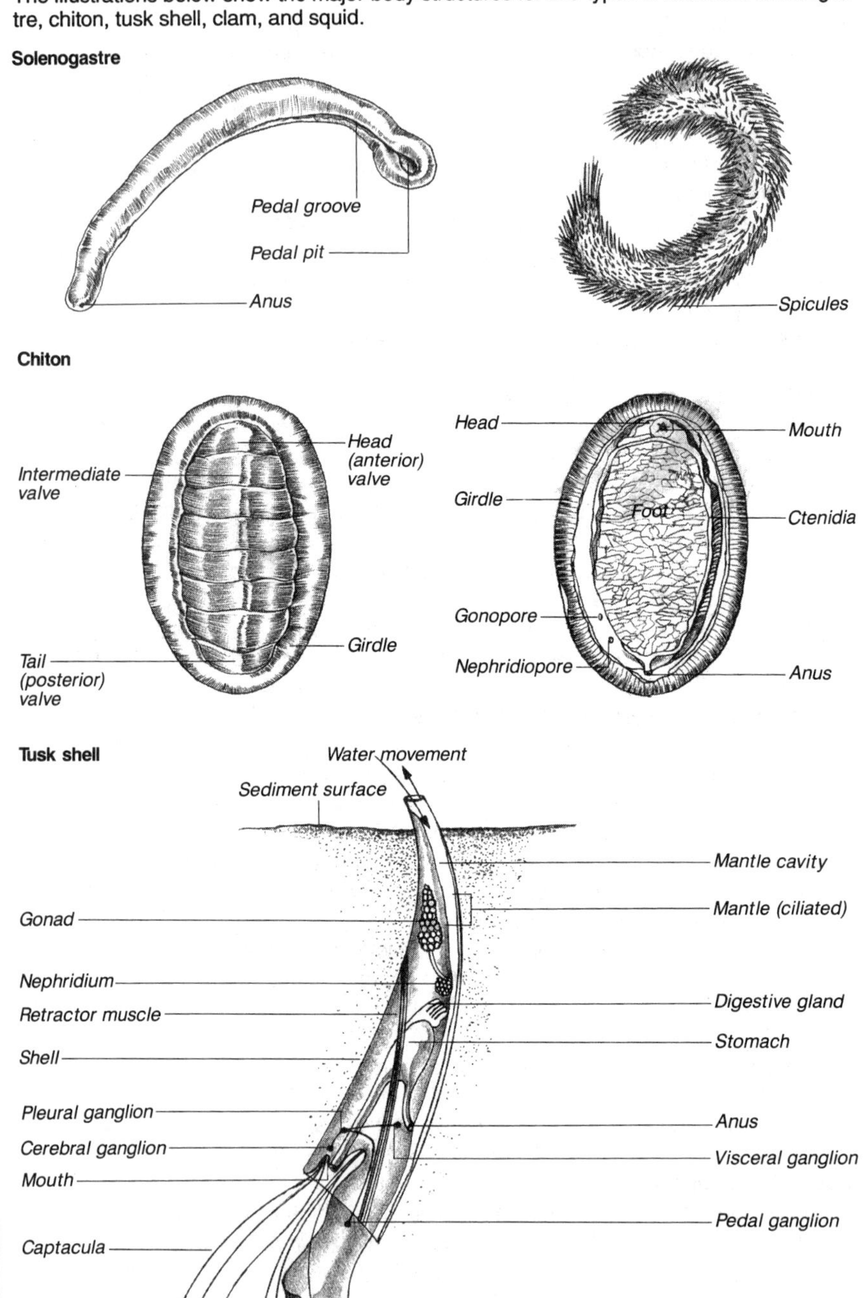

(continued)

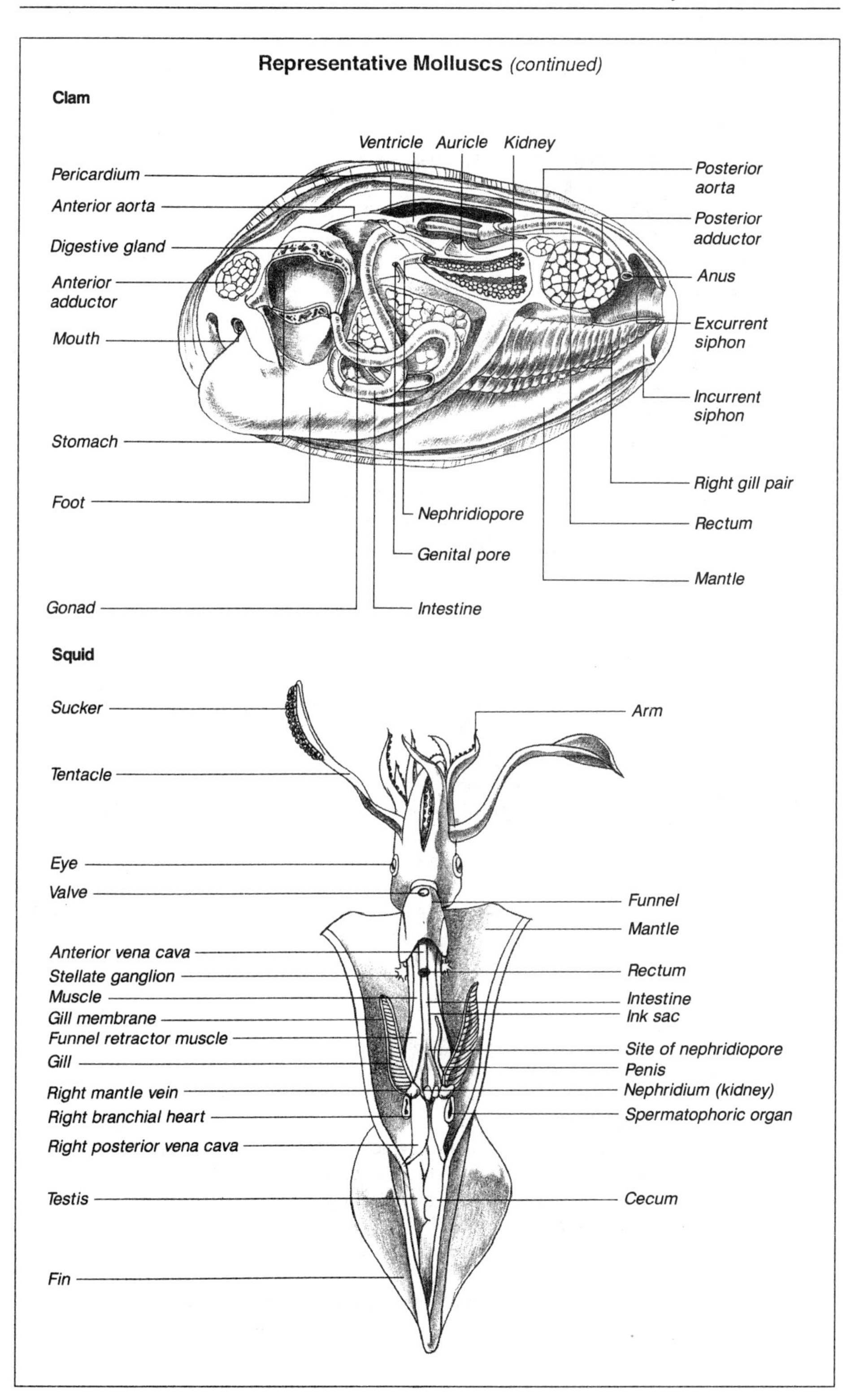
Representative Molluscs (continued)
Clam
Ventricle
Auricle
Kidney
Pericardium
Posterior aorta
Anterior aorta
Posterior adductor
Digestive gland
Anus
Anterior adductor
Excurrent siphon
Mouth
Incurrent siphon
Stomach
Right gill pair
Foot
Nephridiopore
Rectum
Genital pore
Mantle
Gonad
Intestine
Squid
Sucker
Arm
Tentacle
Eye
Valve
Funnel
Mantle
Anterior vena cava
Stellate ganglion
Rectum
Muscle
Intestine
Gill membrane
Ink sac
Funnel retractor muscle
Site of nephridiopore
Gill
Penis
Right mantle vein
Nephridium (kidney)
Right branchial heart
Spermatophoric organ
Right posterior vena cava
Testis
Cecum
Fin

c. The mantle cavity surrounds the foot
d. Internal organs include six to 80 pairs of ctenidia, which are suspended from the mantle cavity, and one pair of metanephridia
e. The head is reduced; cephalic eyes and tentacles are absent, although photoreceptors may be distributed around the shell and mantle surface
f. Chemosensors are abundant

2. Most chitons are small (2 to 5 centimeters in length), although some species grow as large as 30 centimeters
3. They are common inhabitants of rocky intertidal habitats with a few deep sea species; all chitons are marine animals
4. Chitons primarily are herbivorous
    a. They feed on algae, which they scrape from rock surfaces with a radula that has many transverse rows of teeth
    b. A few species are predatory carnivores
5. Chitons are dioecious
    a. Fertilization is external; eggs are shed into the sea or brooded within the mantle cavity
    b. The trochophore larva develops directly into the adult
6. About 600 living species have been described
7. Examples include *Choriplax, Chaetopleura,* and *Cryptochiton*

F. Class Scaphopoda

1. Scaphopods commonly are called *tusk* or *tooth shells*
2. Most species are small (about 2 to 5 centimeters long), but some may grow to 25 centimeters in length (see *Representative Molluscs,* page 82)
3. These molluscs have a tubular shell that tapers at the posterior end
    a. The shell usually is open at both ends
    b. The large mantle cavity extends along the entire ventral surface
    c. The head is reduced, and eyes are absent
    d. Paired arrays of contractile tentacles, called *captacula,* are used to capture prey
    e. Internal organs include a radula and paired metanephridia; the heart is absent
4. About 350 living species of tusk shells have been described; all live in benthic marine habitats
5. Examples include *Fustiaria* and *Cadulus*

G. Class Gastropoda

1. Gastropoda is the largest mollusc class
    a. About 40,000 living species have been described
    b. This class includes the snails and slugs, which range from primitive to highly specialized terrestrial forms
2. Gastropods are found in marine, freshwater, and terrestrial habitats; most are sedentary and slow moving, a consequence of heavy protective shells
3. They have a single coiled or uncoiled shell; slugs, however, do not have shells
4. They have torted, asymmetrical bodies
    a. This body form has been very successful; about 80% of all living molluscs are snails
    b. Some snails can withdraw their bodies into the shell for protection

c. An operculum, a horny plate that covers the shell aperture when the body is withdrawn, may be present

5. Some gastropods experience a reversal of torsion during development (after the veliger larval stage); torted gastropods have structural adaptations to avoid fouling
   a. Restrictions of the asymmetrically coiled shell cause organ reductions on the right side of the body; the right ctenidium, atrium, metanephridium, and osphradium are reduced or absent
   b. Water is brought in on the left side of the mantle cavity and flows out the right side; the excurrent flow removes wastes from the nephridiopore and anus on the right side of the body, avoiding fouling
6. The muscular foot of gastropods may be adapted for locomotion, attachment to the substrate, or burrowing
7. The radula is used for food gathering
8. Gastropods may be dioecious or cross-fertilizing hermaphrodites
   a. Living gastropods have only one gonad, located within the visceral mass
   b. In species with a nonfunctional right metanephridium, right excretory structures are converted for use solely in reproduction
      (1) As a result, complex reproductive structures and developmental patterns have evolved in those species
      (2) These include the addition of egg shells or horny capsules, and, in terrestrial pulmonates, the internal brooding of embryos
   c. During copulation, hermaphroditic species exchange sperm packets, called *spermatophores*
   d. Many species perform precopulatory courtship ceremonies
9. Most gastropods bypass or minimize the trochophore larva development stage
   a. In many groups, embryos hatch as veliger larvae
   b. Many advanced species have direct development; the veliger stage occurs within the egg case
10. The three gastropod subclasses are Prosobranchia, Opisthobranchia, and Pulmonata
   a. Subclass *Prosobranchia* includes most gastropods (more than 13,000 species)
      (1) Prosobranchs primarily are shelled marine snails; a few species live in freshwater habitats
      (2) This subclass is characterized by the presence of an operculum; the shell usually is coiled but may be dome-shaped or tubular
      (3) Examples include limpets, periwinkles, whelks, conchs, and abalones
   b. Subclass *Opisthobranchia* is smaller (about 2,000 species) and includes primarily nonshelled marine gastropods, or sea slugs
      (1) All species are marine
      (2) The group is characterized by various degrees of detorsion (secondary reversal of torsion)
      (3) In some species, such as nudibranchs, the mantle cavity is absent
         (a) Gas exchange occurs by anal gills or by diffusion across the body wall
         (b) Long extensions of the body wall, called *cerrata,* increase the surface area for diffusion
      (4) Examples include nudibranchs, sea hares, sea butterflies (planktonic gastropods), bubble shells, and canoe shells

c. Subclass *Pulmonata* contains terrestrial and freshwater slugs
   (1) Of the 16,000 described species of pulmonates, most are terrestrial
   (2) Gills are absent; the mantle cavity is converted into an air-breathing lung
   (3) The shell usually is present and coiled
   (4) The body may be detorted
   (5) The nervous system is highly cephalized, and the sensory organs, including eyes and one or two pairs of tentacles, are well developed

H. Class Bivalvia (Pelecypoda)

1. Bivalves are molluscs with two laterally compressed shells (see *Representative Molluscs,* page 83)
   a. The shell is hinged dorsally with elastic ligaments
   b. Powerful *adductor muscles* close the shell; scallops clap the shell valves together with giant adductor muscles, producing a type of jet propulsion
   c. The head is reduced; eyes and radula are absent
   d. A large mantle cavity is present, and mantle tissue covers both sides of the body; the mantle often is fused to form *inhalant* and *exhalant siphons*
   e. The foot is laterally compressed
      (1) It may be used for locomotion or burrowing
      (2) In attached bivalves, the foot and anterior end are reduced; in some attached species, such as mussels, the foot secretes adhesive anchoring strands called *byssal threads*
   f. Only one pair of metanephridia are present
2. About 8,000 living species have been described; bivalves are found in marine and freshwater habitats
3. They range in size from less than 1 centimeter to more than 1.3 meters in length, which is the size of the giant clam *(Tridacna)*
4. Most bivalves are filter feeders
5. Most bivalves are dioecious
   a. Fertilization usually is external
   b. The embryos may pass through up to three developmental stages—trochophore, veliger, and glochidium (juveniles)—before adulthood
6. Examples of common bivalves include clams, oysters, mussels, and scallops

I. Class Cephalopoda

1. Cephalopods are the most advanced of the molluscs; these active predators live exclusively in marine habitats
2. About 650 living species have been described; many more are known from the fossil record
3. The head and foot are combined into an anterior structure that is divided into eight to ten prehensile *tentacles* or *arms* (see *Representative Molluscs,* page 83)
   a. In most species, the arms and tentacles are equipped with *suckers*
   b. In the male, one pair of arms is modified for copulation
   c. Elongation along the dorsal-ventral axis is exaggerated
   d. The chitinous, beak-like jaws are used to bite off pieces of prey; these pieces are pulled into the mouth by a modified radula
   e. Except for *Nautilus,* the shell is reduced or absent

f. The nervous system is well developed and complex; it is the most advanced of the molluscs and may be the most advanced of any invertebrate
  (1) Cephalopods are highly cephalized
  (2) Sensory receptors are well developed, including a complex eye with cornea and lens and an elaborate statocyst; vision in some cephalopods is comparable to that of vertebrates
g. The circulatory system, which is well developed and complex, supports the cephalopod's large size and active life-style; flow from the hemocoel to the lungs is enhanced by powerful *branchial hearts*, which boost the low venous pressure
h. ***Chromatophores*** (pigment-containing cells) in the epidermis allow cephalopods to adopt a broad array of colors and patterns; many squid are also *bioluminescent*
i. Most species have an *ink sac* that empties into the rectum
  (1) The sac contains an ink gland that secretes *sepia*, a dark fluid
  (2) Ink release is an alarm response; it may be intended to confuse a predator or to camouflage the cephalopod's escape

4. Cephalopods range in size from 2 to 3 centimeters up to the giant deep sea squid, *Architeuthis*, which can reach 16 meters in length
5. Most cephalopods swim by jet propulsion; water is expelled from the mantle cavity through a ventral funnel
6. Cephalopods are active predators; they feed on fish, other molluscs, crustaceans (see Chapter 11, Phylum Arthropoda), and worms
7. Cephalopods have separate sexes
   a. Elaborate precopulatory rituals may occur
   b. The male uses specialized arms to transfer spermatophores (sperm packets) into the female's mantle cavity
   c. Eggs float or are attached to the substrate; octopus females tend the eggs until they hatch
   d. Development in cephalopods always is direct
8. Cephalopods include the squid, octopus, cuttlefish, and the chambered *Nautilus*

Study Activities

1. List five basic characteristics of molluscs.
2. Explain current hypotheses concerning the evolutionary ancestry of molluscs.
3. Describe how molluscs carry out nutrition, gas exchange, and excretion.
4. Describe methods of reproduction in molluscs.
5. Create a chart that compares and contrasts the basic characteristics of each of the eight mollusc classes.

10

Phylum Annelida

Objectives

After studying this chapter, the reader should be able to:
- Describe the basic characteristics of annelids.
- Explain current hypotheses concerning the evolutionary ancestry of annelids.
- Describe how annelids perform their basic life functions.
- Identify and characterize the classes of phylum Annelida.

I. Basic Characteristics

A. General information

1. Members of the phylum Annelida, which includes the segmented worms, have elongated, segmented bodies with chitinous bristles
2. Annelids are eucoelomate protostomes, with a closed circulatory system and a well-developed nervous system
3. As with several other protostomes, many annelids have a trochophore larval stage during development
4. Familiar examples include earthworms and leeches

B. Ecologic relationships

1. Annelids are found in marine, freshwater, and terrestrial habitats
2. This relatively large phylum includes about 15,000 species
3. Annelids are diverse in ecology and behavior; they may be free-swimming or benthic, and many live in burrows or tubes
4. Symbiotic (parasitic, mutualistic, or commensal) relationships are common among annelids
5. Free-living annelids may be sediment feeders, filter feeders, herbivores, or predatory carnivores
6. Annelids are important members of aquatic and terrestrial food chains and thus have indirect economic importance
7. Burrowing worms aerate and mix the soil of both terrestrial habitats and aquatic sediments
    a. The numerous deposit-feeding worms may pass thousands of tons of sediments through their guts each year
    b. This has an important effect on the composition of the substrate

C. Evolutionary relationships

1. Annelids, molluscs, and arthropods may have arisen from a common flatworm ancestor
2. The partitioned bodies of annelids and arthropods evolved from a nonsegmented coelomate ancestor, probably a burrowing form
3. Scientists disagree over whether the oligochaete or polychaete body plan arose first; leeches are closely related to the oligochaetes and probably are an offshoot of that line

II. Annelid Form and Function

A. General information

1. Annelids have bilaterally symmetrical, elongated bodies
   a. The body is divided into a longitudinal series of repeating segments called *metameres*
   b. Only the body trunk is segmented; the head and anal region are not considered to be metameres
   c. The body has chitinous bristles called *setae* (except in leeches)
2. The coelom is well developed and divided by septa (except in leeches)

B. Body types of annelids

1. The body usually is cylindrical in cross section but may be dorsoventrally flattened (as in leeches)
2. Segments are divided externally by grooves or serial repetition of appendages (see *Anterior View of Two Annelids*, page 90)

III. Internal Structure and Physiology of Annelids

A. General information

1. In annelids, the body wall has strong longitudinal and circular muscle layers, and the epidermis is covered with a thin, nonchitinous cuticle
2. Segmentation is reflected internally by the metameric arrangement of tissues and organs (see *Anterior View of Two Annelids*, page 90)
   a. In primitive annelids, each body segment contains a portion of the nervous, circulatory, and excretory systems
   b. Other annelids have groups of segments that are specialized for different functions
3. The body form of different annelid species reflects their life-style
   a. Active forms usually have similar segments
   b. Sedentary forms have greater regional segmental specialization
   c. Specialized structures include tentacles, fleshy appendages (called *parapodia*), and feathery crowns of ciliated arms (as in feather duster worms and Christmas tree worms)
4. The coelomic fluid functions as a hydrostatic skeleton; the internally partitioned bodies of annelids facilitates the control of hydrostatic pressure for muscle contraction
5. The digestive system is complete with regional functional specialization

Anterior View of Two Annelids

These illustrations show dorsal views of the internal and external body structures for the polychaete *Nereis* and the oligochaete *Lumbricus*. Segments of *Nereis* are divided externally by grooves; note the serial repetition of appendages and nephridia. By contrast, *Lumbricus* has conspicuous internal segmentation and a less developed head, and lacks parapodia.

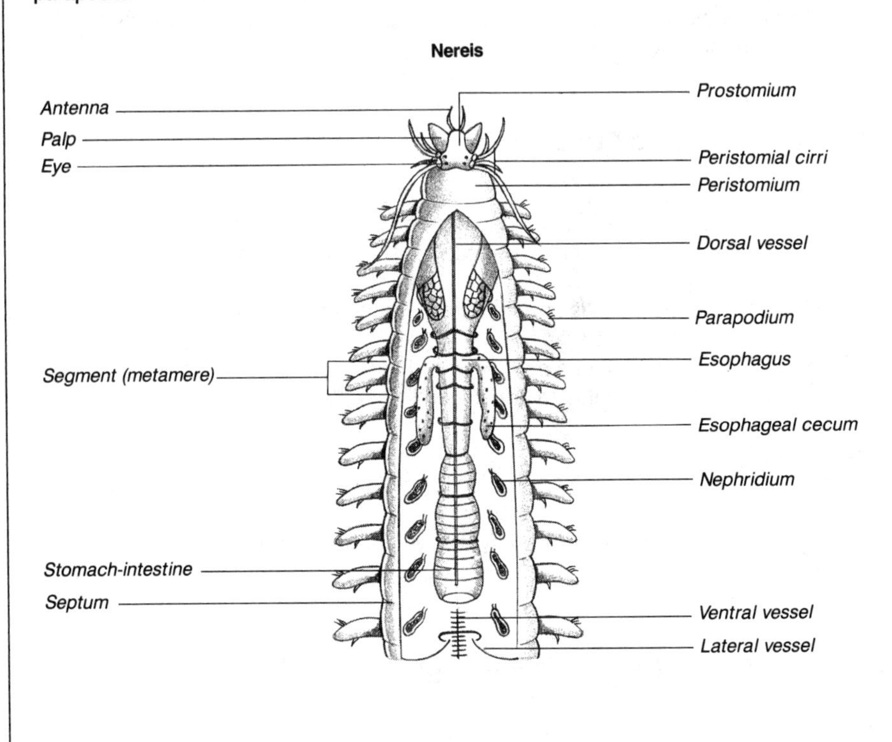

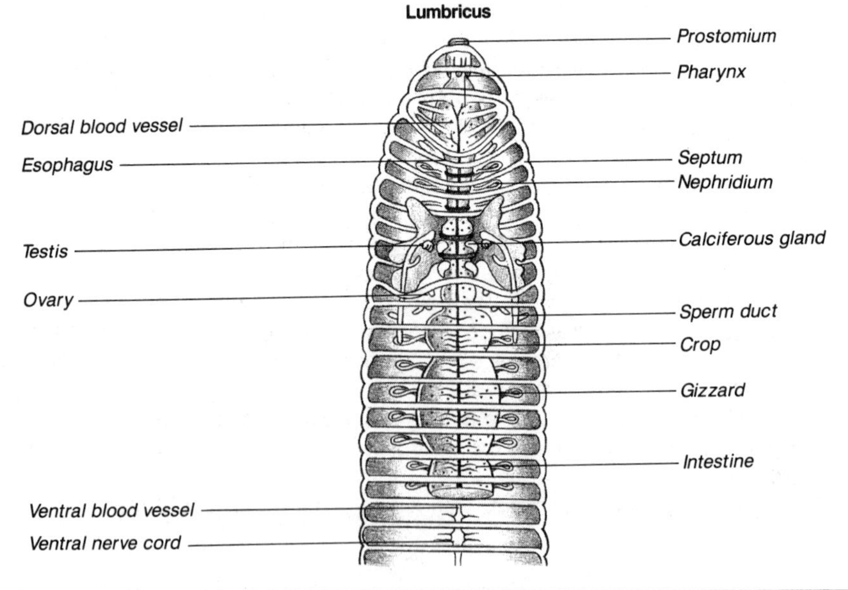

6. Excretory organs typically consist of metanephridia but may be protonephridia; each segment usually contains one pair of metanephridia
7. The nervous system and sensory receptors are well developed

B. Locomotion

1. Annelid bodies are adapted for crawling, burrowing, and swimming
2. Longitudinal muscle layers provide the main propulsive force; circular muscle layers maintain and distribute the hydrostatic pressure among the metameres
3. Some worms use their muscular parapodia for walking; in burrowing forms, the parapodia are reduced or absent
4. Leeches have anterior and posterior suckers to anchor the body and facilitate muscle contractions

C. Nutrition

1. Annelid worms may be predatory carnivores, deposit feeders, filter feeders, or ectoparasites
2. *Predatory carnivores* feed primarily on small invertebrates
   a. Some annelid predators can move rapidly across the substrate to capture prey
   b. Others live in tubes or burrows and detect the presence of prey with chemosensors or vibration sensors
      (1) These annelids use a protrusible proboscis to harpoon or capture prey outside the burrow
      (2) In some cases, the jaws have poison glands
   c. Still others are ambush predators, seizing prey as they pass by
3. *Deposit feeders* burrow through the substrate, ingesting sediments as they move forward; they may be selective or nonselective feeders
4. *Filter feeders* have elaborate feeding structures, such as crowns of ciliated tentacles, that trap organic particles in the water
   a. Cilia and the mucus on feeding tentacles trap food particles and carry them to the mouth
   b. Many tube-dwelling annelids are filter feeders
5. *Ectoparasites* feed on the blood and body fluids of the host
   a. Most leeches (class Hirudinea) are parasites of vertebrates, but a few feed on invertebrates (including other annelids, and even other leeches)
   b. Leeches attach to the host with their suckers, make an incision with blade-like jaws, and suck blood or body fluids with their muscular pharynx
6. The annelid digestive tract is unsegmented and can be divided into three basic regions—the foregut, the midgut, and the hindgut
   a. The *foregut* functions primarily in ingestion, transport, storage, and mechanical digestion of food
      (1) Food taken in through the pharynx passes into the esophagus
      (2) The esophagus may be modified into specialized regions
         (a) The *crop* functions in food storage
         (b) One or more *gizzards* mechanically grind food into small particles
         (c) *Calciferous glands,* if present, remove excess calcium from ingested material and precipitate it (as calcite) into the intestine for excretion; the glands also may regulate the concentration of calcium and carbonate ions in body fluids
   b. The *midgut* is composed of a long, straight intestine

(1) The intestine is the site of extracellular digestion and absorption
(a) Secretion of digestive enzymes into the lumen of the intestine occurs in the anterior region
(b) Absorption occurs primarily in the posterior region of the intestine
(2) In terrestrial species, the intestinal surface area is greatly increased by a dorsal longitudinal fold
c. A layer of pigmented cells, called *chloragogenous tissue,* surrounds the midgut
(1) This tissue plays an important role in carbohydrate, lipid, and protein metabolism
(2) In terrestrial species, the chloragogenous tissue also may function in urea production
d. The *hindgut* functions in excretion

D. Reproduction
1. Annelids reproduce both asexually and sexually
2. Asexual reproduction is found only among oligochaetes and polychaetes; hirudineans are not capable of asexual reproduction or regeneration
a. Some annelids reproduce asexually by fragmentation; the worms separate their bodies into individual segments, each of which regenerates a new individual
b. Other annelids reproduce asexually by budding; each bud produced on the parent's body develops into a new individual
3. Annelids that reproduce sexually may be dioecious or hermaphroditic
4. Most polychaetes are dioecious, although a few are hermaphroditic
a. Polychaetes do not have permanent gonads or accessory structures
b. Gametes are produced seasonally from peritoneal swellings in various body regions and are shed into the coelom
c. Fertilization is external; gametes exit the body through gonoducts, the nephridiopore, or by rupture of the body wall
d. Development in polychaetes is indirect, with a trochophore larval stage
5. Oligochaetes are cross-fertilizing hermaphrodites
a. They possess complex, permanent gonads and reproductive structures (see *Anterior View of Two Annelids*, page 90)
(1) The gonads are found only in specific segments, usually in the anterior part of the body
(2) The male organs include one or two pairs of testes
(a) Sperm are released into the coelom
(b) They may mature in the coelom or be stored in seminal vesicles during maturation
(c) When mature, the sperm are carried by sperm ducts to the gonopores
(3) A single pair of ovaries is located behind the male organs
(a) Eggs are released into the coelom
(b) As with sperm, they may mature in the coelom or be stored in pouch-like ovisacs during maturation
(c) Eggs are released through the female gonopore
(4) Most oligochaetes have pairs of seminal receptacles that store sperm after copulation

(5) Fertilization is internal; sperm travel along seminal grooves to the seminal receptacle of the opposite worm

b. A region of glandular tissue called the *clitellum* plays an important role in reproduction

(1) Mucus secreted from this area holds the worms in position during copulation

(2) The clitellum also secretes a protective protein casing, called the *cocoon,* for the egg capsule; the cocoon is tough and resistant to adverse environmental conditions

(3) Eggs and sperm are released into a thick layer of albumin secreted by the clitellum; fertilization occurs within this albumin matrix

c. Development in oligochaetes is direct; there is no larval stage

6. Hirudineans (leeches) also are cross-fertilizing hermaphrodites

a. They have complex reproductive organs (see *Structure of a Leech*); the clitellum is present only during the breeding season

b. Fertilization is internal; sperm are transferred by the penis or by hypodermic injection through the body wall

c. Cocoon formation and development are similar to that of oligochaetes

E. Circulation and gas exchange

1. In most annelids, gas exchange occurs by diffusion across the body wall

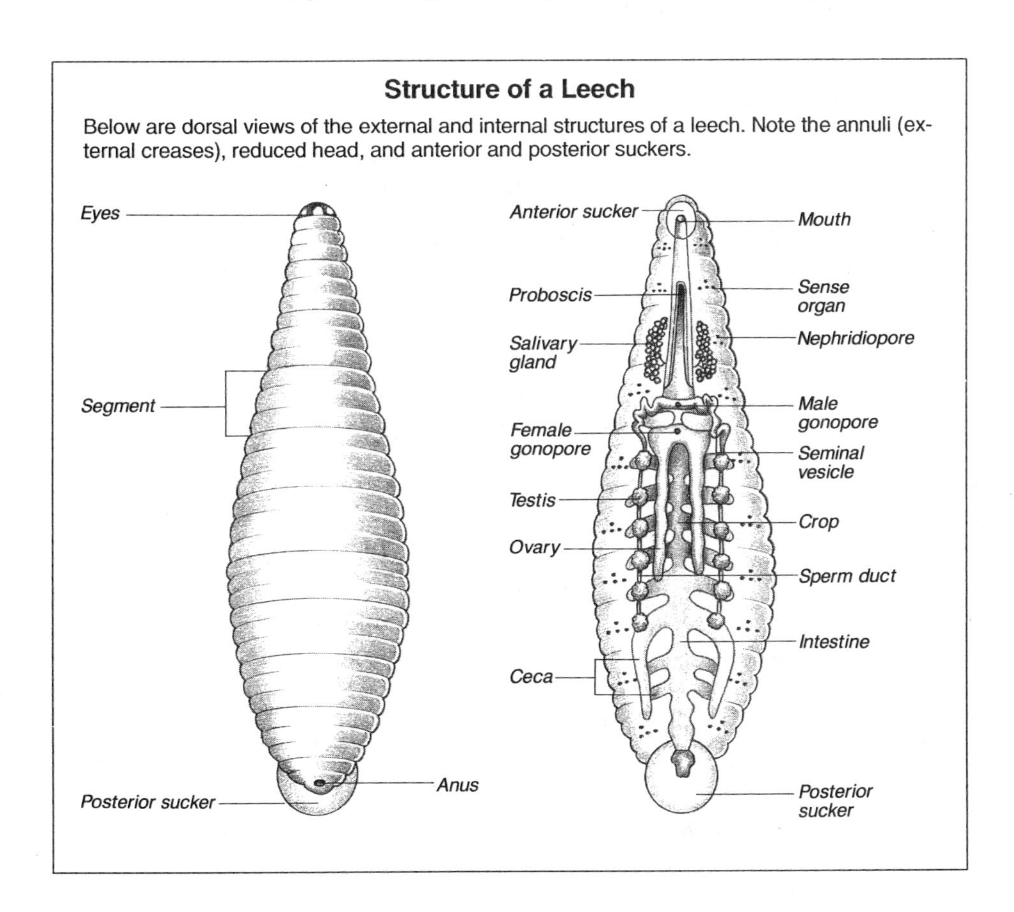

Structure of a Leech

Below are dorsal views of the external and internal structures of a leech. Note the annuli (external creases), reduced head, and anterior and posterior suckers.

a. Terrestrial oligochaetes remove oxygen from the air; if submerged in water, the worms drown because sufficient gas exchange cannot occur
b. Some polychaetes use highly vascularized areas of the parapodia as functional gills
2. Annelids have an efficient circulatory system that functions in the transport of gases, nutrients, and wastes; the coelomic fluid also functions in circulation
3. Most annelids have a closed circulatory system
a. Several longitudinal blood vessels run the length of the body
b. Blood flows posteriorly in ventral blood vessels and anteriorly in the dorsal blood vessel
4. From two to five pairs of hearts pump blood from the dorsal to the ventral vessels
a. These hearts are enlarged muscular regions of the dorsal and other longitudinal blood vessels
b. They usually are located in the esophageal region
5. Respiratory pigments are dissolved in the blood plasma
a. The type of respiratory pigment is related to the activity level and habitat of the annelid
b. An individual worm may have more than one type of respiratory pigment, including alternate forms of hemoglobin
c. Phagocytic amoebocytes are present in the circulatory fluids of most annelids

F. Excretion and osmoregulation

1. The primary organs of excretion and osmoregulation are metanephridia; a few species have protonephridia
2. Generally, annelids have one pair of nephridia per segment
a. Each nephridium occupies parts of two metameres
b. The nephrostome (the entry point for fluid-containing wastes) is located in the segment anterior to the main body of the nephridium
c. The main body of the nephridium is composed of highly vascularized loops; reabsorption of materials occurs in these loops
d. The remaining wastes continue through the nephridial loops to the nephridiopore, where they are released
3. The primary waste product of aquatic species is ammonia; terrestrial forms typically excrete urea
4. Chloragogen cells (part of the chloragogenous tissue) may function in waste removal, but the exact mechanism is not known

G. Nervous system and sense organs

1. The nervous system of annelids is similar in basic structure to the ladder-like system of flatworms, but it is more centralized and has fewer longitudinal nerve cords
2. The basic nervous system plan consists of a pair of dorsal cerebral ganglia at the anterior end and one or more ventral longitudinal nerve cords that run the length of the body; the ganglia are connected to the longitudinal nerve cords by a connective ring that encircles the foregut region
3. Annelids have well-developed sensory receptors
a. Polychaetes have a variety of tactile, photoreceptors, and chemoreceptors distributed over the body surface; most burrowing and tube-dwelling forms have statocysts and georeceptors that enable them to orient themselves in the substrate

b. Oligochaetes have epithelial sense organs distributed all over the body; most of these sense organs are extremely sensitive tactile and chemoreceptors
c. Hirudineans have the least complex sensory receptors
 (1) They have a large array of epidermal sense organs, similar to those of the oligochaetes, including vibration and tactile receptors
 (2) They may have two to ten simple eyes; most leeches are *negatively phototactic,* that is, they tend to avoid light

IV. Classification of Phylum Annelida

A. General information

1. The three classes in phylum Annelida are Polychaeta, Oligochaeta, and Hirudinea
2. The classes differ in appearance, mode of reproduction, and life-style

B. Class Polychaeta

1. The class Polychaeta includes the most primitive of annelids
2. Polychaetes exhibit a great diversity of structural forms
 a. *Errant polychaetes* are adapted for benthic, crawling life-styles
  (1) The head and sensory structures are well developed
  (2) The large parapodia function like legs for crawling
  (3) Large, well-developed teeth and jaws are used to capture prey; most errant species are predators
 b. *Gallery-dwelling polychaetes* are adapted for burrowing in sand or mud; they may construct extensive tunnel systems
  (1) The head and sensory structures usually are simple, except for those burrow dwellers that leave their burrows to feed as predators (such as *Nereis*)
  (2) Because most gallery-dwelling polychaetes move through their burrow system by muscular contractions of the body wall, parapodia are reduced in size
  (3) Many species are sit-and-wait predators that sense nearby prey by vibrations near the burrow opening; an armed protrusible proboscis may be used to capture prey
 c. *Sedentary burrowers* are adapted for a sessile life-style
  (1) Most of these species construct simple vertical burrows with few outside openings
  (2) Because sedentary burrowers seldom move, the parapodia frequently are modified into hook-like ridges used for anchoring to the burrow walls
  (3) Most species are deposit feeders; they may have specialized feeding structures, such as tentacles, that remove organic matter from the substrate
 d. *Tube-dwelling polychaetes* secrete protective tubes; most polychaetes are tube dwellers
  (1) Some species are predators and have adaptations for rapid movement and prey capture (similar to those of errant forms)
  (2) Most are sedentary filter feeders; few sensory receptors are present in the head area

(a) As with burrowing forms, parapodia typically are modified into hook-like ridges, which are used for anchoring to the burrow walls
(b) Feeding structures are elaborate, such as the beautiful feather-like radioles of the feather duster and fan worms
(c) In one group (*Chaetopteridae*), parapodia are modified into fans, which produce a powerful water current through the secreted tube; a mucous bag removes food particles from the water column

3. Most polychaetes are marine animals and are found from intertidal habitats to deep-sea trenches
4. They range in length from less than 1 millimeter to more than 3 meters; they may display a variety of beautiful colors and patterns
5. The distinct and well-developed head bears a variety of sense organs, including eyes and tentacles (see *Anterior View of Two Annelids,* page 90)
6. Each segment usually has a pair of fleshy parapodia with bundles of setae; parapodia may be highly modified in structure and function in gas exchange (as gills), locomotion, and food gathering
7. Polychaetes are dioecious with seasonal gonad formation
   a. Reproductive structures are simple, and the clitellum is absent
   b. Development is indirect, with a trochophore larval stage

C. Class Oligochaeta

1. Most oligochaetes live in freshwater or terrestrial habitats; however, many species now are being discovered in marine environments
2. Because they have few structural adaptations to avoid desiccation, terrestrial forms usually are found in moist habitats
   a. Gas exchange in terrestrial species occurs by diffusion across the body wall
   b. The body wall is highly vascularized to facilitate diffusion and transport of oxygen
   c. Since air has a high oxygen concentration, oxygen transfer is adequate even for large terrestrial species
   d. Mucus-producing glands in the epidermis and the release of coelomic fluid through dorsal pores keep the body moist enough for diffusion to occur
3. Oligochaetes range in length from less than 1 millimeter to more than 3 meters
4. The distinctly segmented body has no parapodia, and the number of segments varies
   a. Setae are present but sparsely distributed over the body surface
   b. The head is not distinct; head appendages are few in number and less well developed than those in polychaetes
5. Oligochaetes are hermaphrodites
   a. They have complex, well-developed reproductive structures
   b. Development is direct, with no larval form

D. Class Hirudinea

1. Hirudineans are found in marine, freshwater, and terrestrial habitats
2. Most are 2 to 6 cm long, but a few species can grow to 25 cm in length
3. The class Hirudinea has two subclasses (*Branchiobdellida* and *Acanthobdellida*); these small worms have characteristics that are intermediate between oligochaetes and hirudineans
4. Most members of this class are leeches
   a. Leeches are dorsoventrally flattened, although not as much as flatworms

b. They may exhibit a variety of colors and patterns

5. Leeches have a fixed number of segments (usually 34); however, external creases (called *annuli*) make them appear as though they have numerous smaller segments (see *Structure of a Leech,* page 93)
6. Leeches are characterized by anterior and posterior suckers
   a. Setae and parapodia are absent
   b. The coelom is reduced and does not have internal septa
      (1) The coelomic spaces are filled with connective tissue and muscle
      (2) The coelomic fluid does not function as a hydrostatic skeleton
7. Leeches are hermaphrodites
   a. Reproductive structures are complex and well developed
   b. Development is direct, without a larval stage
8. Leeches display a variety of feeding habits, from predatory carnivores to temporary or permanent ectoparasites
   a. They are highly sensitive to the body secretions and odors of a potential prey or host
   b. Those that feed on mammals also are attracted by body warmth
   c. Predatory leeches feed frequently, but ectoparasitic forms consume large meals (up to several times their body weight)
      (1) Because leeches have few digestive enzymes, they digest their food slowly; they may take several months to digest a single blood meal
      (2) Symbiotic bacteria within the intestinal tract may be the primary agents of digestion
   d. Leeches apparently are not affected by long periods of fasting (depending on host availability); some species can survive more than a year without feeding

Study Activities

1. List four basic characteristics of the annelids.
2. Explain current hypotheses concerning the evolutionary ancestry of annelids.
3. Describe how annelids carry out nutrition, gas exchange, and excretion.
4. Describe methods of sexual and asexual reproduction in annelids.
5. Create a chart that compares and contrasts the basic characteristics of each of the three annelid classes.

11

Phylum Arthropoda

Objectives

After studying this chapter, the reader should be able to:
- Describe the basic characteristics of arthropods.
- Explain current hypotheses concerning the evolutionary ancestry of arthropods.
- Describe how arthropods perform their basic life functions.
- Identify and characterize the subphyla, classes, and orders of phylum Arthropoda.

I. Basic Characteristics

A. General information

1. Arthropods are the largest and most diverse group of animals on earth; phylum Arthropoda includes the insects, spiders, scorpions, ticks, mites, crustaceans, millipedes, centipedes, and many extinct species, such as the trilobites
2. They are characterized by a hard exoskeleton and jointed appendages (arthropod means "jointed foot")
3. These eucoelomate protostomes have well-developed organ systems
4. Arthropods and annelids have many similarities and share close evolutionary ties

B. Ecologic relationships

1. Arthropods are found in every known habitat on earth
2. They range in size from parasitic mites that are less than 0.1 millimeter long to the Japanese spider crab, which has a leg span of more than 4 meters
3. Because of their great diversity, large numbers of species, and wide ecologic distribution, arthropods are considered to be the most successful animal phylum
4. Almost 1 million species of arthropods have been described; some specialists believe that over 50 million additional species remain undiscovered, especially in tropical regions
5. Several structural and physiologic features have contributed to the remarkable success of this group
   a. The arthropod exoskeleton (a chitinous cuticle) provides a high degree of protection without restricting mobility
   b. Segmentation and a variety of specialized, jointed appendages facilitate food gathering, locomotion, and adaptation to diverse habitats
   c. A highly efficient oxygen transfer system supports an active life-style
   d. The nervous system and sensory organs are well developed; most arthropods are capable of complex, learned behaviors

e. Development through metamorphosis permits larval forms to exploit habitats different from those of the adult
f. The ability to fly has led to wide geographic distribution

C. Evolutionary relationships

1. Arthropod evolution is a controversial issue, with several differing viewpoints; the problem is magnified by a sparse fossil record of primitive arthropods and proposed ancestral forms
2. Although the exact lineage is under dispute, scientists agree on several points about arthropod ancestry
   a. There is a clear link between arthropods and annelids
      (1) The body plans of the two groups are similar
      (2) Both groups are segmented, with repetitive, paired, external and internal structures
      (3) The structure of the brain and other nervous system features is similar
      (4) There are similar patterns of growth and development
   b. There is a clear link between this annelid-arthropod line and two other phyla, the onychophorans and the tardigrades
      (1) *Onychophorans* are an early group of worm-like animals
         (a) Like the annelids, they have a body wall and metanephridia
         (b) Like the arthropods, they have a reduced coelom, open circulatory system, and tracheal breathing system
         (c) The cuticle contains chitin
         (d) There are overall similarities in appearance to arthropods
      (2) *Tardigrades* (commonly known as water bears) are microscopic animals
         (a) They are similar in appearance to arthropods
         (b) They have a reduced coelom and an open circulatory system
         (c) The cuticle contains chitin
3. The exact degree of relationship among the annelids, arthropods, onychophorans, and tardigrades is still under discussion
   a. The onychophorans, tardigrades, and arthropods share a common body plan, which is a modification of the basic annelid body plan
   b. Differences include reduction of the coelom and its replacement with the hemocoel as the main body cavity
4. Available evidence suggests that there was an annelid-like ancestor (which may have been polychaete-like, oligochaete-like, or some other form) with ventrally located appendages (rather than the laterally located parapodia of polychaetes)
   a. This ancestral animal gave rise to the onychophorans
   b. The onychophorans led to a second line with some arthropod-like characteristics
   c. This second line gave rise to both the arthropods and the tardigrades

II. Arthropod Form and Function

A. General information

1. Arthropods are eucoelomate protostomes with bilateral symmetry
2. The body is internally and externally segmented; each body segment may bear a pair of jointed appendages

3. The body consists of specialized groups of segments called *tagmata;* examples include the head, thorax, and abdomen
4. A chitinous exoskeleton, divided into jointed plates, covers the body

B. Body types of arthropods

1. Arthropods are characterized by the presence of a hard, jointed exoskeleton that protects and supports the body
2. The exoskeleton is a series of plates and joints that allow movement of the body and appendages
3. The body wall is a multi-layered cuticle (the exoskeleton) secreted by the epidermis; cuticle composition differs among arthropod groups but has the same general structure
   a. The outermost protective layer, sometimes called the *cement layer,* is composed of lipoprotein; it seals the exoskeleton and protects against bacterial penetration
   b. The next layer, called the *waxy layer,* varies in thickness; it protects against water loss and provides a further barrier against bacteria
   c. The *cuticulin layer* produces secretions needed for hardening the exoskeleton
   d. The thick *procuticle* is composed of layers of protein and chitin; it is the main structural layer of the exoskeleton
   e. In some species, the exoskeleton is *sclerotized* (tanned and hardened) or *mineralized* (impregnated with mineral deposits such as calcium carbonate); these processes add rigidity
4. The exoskeleton is nonliving and does not grow with the organism; it is periodically shed to allow growth, a process known as *molting*
5. The jointed appendages of arthropods are outgrowths of the body wall
   a. Short bands of muscle, attached to the inside of the exoskeleton, move the appendages
   b. This is in contrast to the musculoskeletal system of vertebrates, in which the muscles are attached to the external surface of bones
6. Arthropods have only striated muscle fibers; smooth muscle is absent

III. Internal Structure and Physiology of Arthropods

A. General information

1. The arthropod body cavity is an open hemocoel; the hemolymph supports and surrounds the internal organs
2. The gut is complete with regional specialization
3. The nervous system is complex with well-developed sense organs
4. The exoskeleton is molted for growth to occur

B. Locomotion

1. Arthropod limbs are adapted for diverse modes of locomotion
2. Specialized appendages allow various arthropods to crawl, burrow, swim, fly, jump, and make many other movements
3. Many arthropods can voluntarily detach appendages, a process called *autotomy,* which may be used as an escape mechanism when threatened by a predator

C. Nutrition

1. Arthropods have varied diets and feeding methods
2. Several pairs of appendages in the general vicinity of the mouth are specialized for various aspects of food gathering and handling
   a. Large *pedipalps* are adapted for prey capture
   b. In ectoparasites, *chelicerae* are adapted for penetrating and anchoring in the host's epidermis
   c. In filter-feeding species, appendages with many fine, hair-like structures trap food particles
   d. Other appendages are adapted for scraping, tearing, cutting, or crushing
3. Arthropods have a complete digestive tract with regional specializations
   a. The *foregut* region is involved in ingestion, storage, and mechanical breakdown of food; many species have a crop for food storage and a gizzard for food grinding
   b. The *midgut* region is the site of enzyme production, chemical digestion, and absorption; this region often includes one or more *digestive ceca* (also called *digestive glands, liver,* or *hepatopancreas*)
   c. The *hindgut* region is the site of water reabsorption and feces production
4. Several groups of terrestrial arthropods have special glands associated with the gut that produce noxious chemicals to repel predators; other glands secrete various types of silk webs or cocoons

D. Reproduction

1. Most arthropods are dioecious and have a single pair of gonads and gonoducts
2. Most species participate in some type of formal mating behavior
3. Fertilization usually is internal but may be external in some aquatic species
4. Development may be direct or indirect, depending on the species
5. Parental care usually is present, especially during early embryologic development
6. Because the arthropod skeleton does not grow with the organism, it is periodically molted (also called *ecdysis*)
   a. The molting process is hormonally controlled
   b. Production of *ecdysone* (the molting hormone) signals the onset of molting

E. Circulation and gas exchange

1. Arthropods have an open circulatory system with a dorsal heart
   a. A large *pericardial sinus* (a cavity that is not part of the coelomic space) surrounds the heart
   b. Blood enters the heart through perforations in the muscular wall, called *ostia*
   c. When the heart contracts, the ostia close, and blood is propelled out of the heart into the arterial system
   d. Hemolymph (blood) flows to the various blood sinuses that make up the hemocoel, where gas exchange with the tissues takes place
2. The complexity of the circulatory system is related, in large part, to body size and shape; larger animals tend to have more complex circulatory structures
3. Arthropod hemolymph transports nutrients, wastes, and gases
   a. In small species, gases dissolve directly in the hemolymph fluids
   b. Larger animals have respiratory pigments; hemocyanin is the most common, but in some species hemoglobin also is present
4. Gas exchange organs differ in aquatic and terrestrial arthropods
   a. Aquatic forms, such as crustaceans, have branched or folded gills

(1) Gills provide an efficient gas exchange mechanism; many aquatic arthropods are quite active
(2) In some species, the gills are protected by the exoskeleton in special respiratory chambers; in others, the gills are exposed

b. Terrestrial species, such as insects and some arachnids, have a complex network of respiratory tubules called *tracheae*
(1) Tracheal tubes branch throughout the body tissues
(2) The inner ends of the tubules open directly to the tissues, allowing direct gas exchange between the air and the internal organs
(a) In species with tracheal systems, blood is not important for gas transport
(b) Efficient gas transfer through the tracheal tubes circumvents the sluggish transport of an open circulatory system
(c) Rapid oxygen transfer at the tissue level facilitates vigorous muscular action, permitting the active life-style of many insects (such as the capability for flight)
(3) Valved openings called *spiracles* connect with the outside air

c. Arachnids and other arthropods, such as horseshoe crabs, may possess *book gills* composed of thin, flat plates called *lamellae;* the gill lamellae may be protected by an exoskeletal flap

F. Excretion and osmoregulation

1. Marine arthropods primarily are osmoconformers and have limited osmoregulatory capabilities
2. Freshwater forms primarily are osmoregulators; they have several mechanisms to maintain their body fluid composition
   a. The exoskeleton is relatively impermeable to water and salts
   b. Rectal glands reabsorb valuable salts and ions from the urine before excretion
   c. Active transport mechanisms import ions from the surrounding water
3. Terrestrial species must conserve water
   a. The impermeable cuticle minimizes dehydration
   b. Excretory products are dry and highly concentrated
4. Aquatic forms excrete most metabolic wastes in the form of ammonia; terrestrial forms excrete primarily water-conserving uric acid
5. In some arthropods, nephridia are the organs of excretion and osmoregulation
   a. Crustaceans have a single pair of nephridia (called *antennal glands* or *green glands*), which are located in the head
   b. Up to four pairs of nephridia (called *coxal glands*) are found in some arachnids
6. Terrestrial species (such as insects and many arachnids) have a tubular excretory system known as *Malpighian tubules*
   a. Water, wastes, and ions are secreted into the tubules
   b. Specialized rectal glands reabsorb the water and ions, producing a relatively dry combination of uric acid and feces for excretion

G. Nervous system and sense organs

1. The general plan of the arthropod nervous system is similar to that found in annelids

a. Two or three separate, but closely connected, dorsal ganglia (brain) are located in the anterior head region
b. The ventral nerve cord may be single or double, with segmental ganglia; a connecting nerve cord encircles the esophagus (see *Internal Structure of a Male Crayfish,* page 104)
2. Sensory receptors are complex and well developed
a. Most are modifications of the exoskeleton, which would otherwise block all sensory input
b. Extensions of the cuticle function as tactile and chemical receptors; statocysts and sound receptors also are present
c. Arthropod eyes may be simple ocelli (with or without a lens) or compound eyes with multiple ommatidia (photoreceptor units)
(1) Some insects may have as many as 30,000 ommatidia in their compound eyes
(2) Most species are have either light- or dark-adapted eyes and lack the ability to adjust their vision to different light conditions

IV. Classification of Phylum Arthropoda

A. General information

1. The arthropods are a large and diverse group of animals
2. The various subphyla and classes show distinct morphologic and physiologic specializations for their diverse life-styles

B. Methods of arthropod classification

1. Some disagreement remains over the correct classification scheme for arthropods, especially below the subphylum level
2. The classification discussed in the following sections is one of the most commonly accepted models
3. An alternative classification scheme differs primarily at the subphylum, class, and subclass levels
a. In this alternative scheme, Merostomata and Arachnida are subclasses within the class Chelicerata
b. Ostracoda, Copepoda, and Cirripedia are subclasses within class Maxillopoda
c. Chilopoda and Diplopoda are subclasses within class Myriapoda

V. Subphylum Chelicerata

A. General information

1. Members of this subphylum are characterized by the absence of antennae; classes include Merostomata (the horseshoe crabs) and Arachnida (spiders, scorpions, ticks, and mites)
2. The two main regions (tagmata) of the body are the cephalothorax and the abdomen
a. The cephalothorax consists of several trunk segments that have fused with the head
b. It often is covered with a single, dorsal exoskeletal plate, called a *carapace*

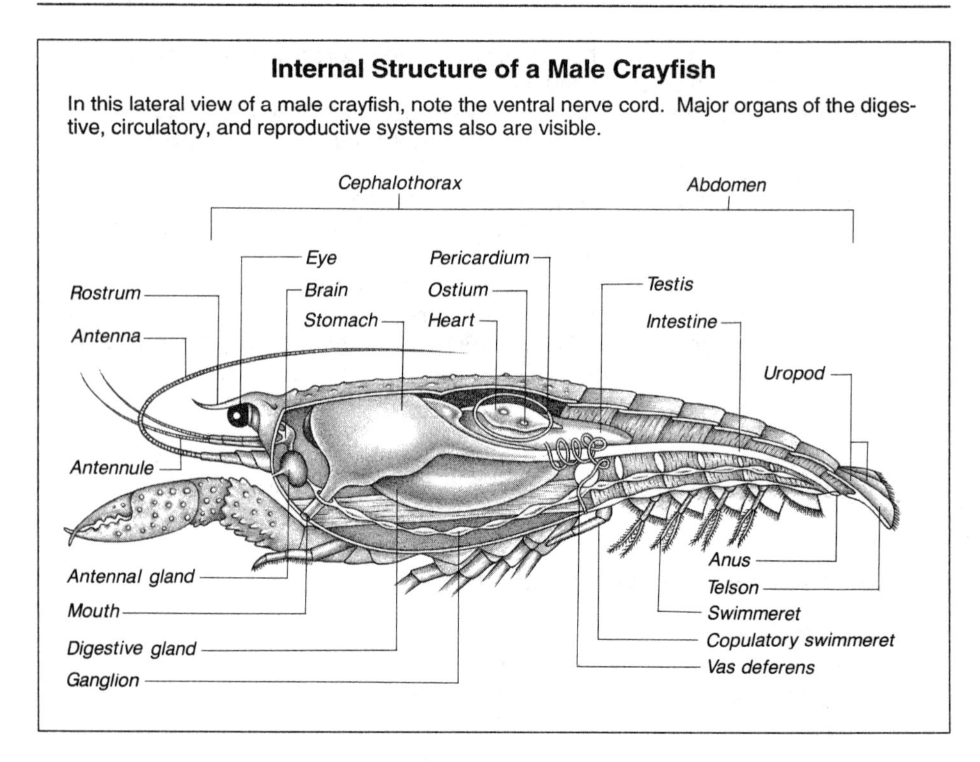

Internal Structure of a Male Crayfish

In this lateral view of a male crayfish, note the ventral nerve cord. Major organs of the digestive, circulatory, and reproductive systems also are visible.

3. The first pair of appendages behind the mouth are called *chelicerae,* which give the subphylum its name; a pair of manipulatory appendages, called *pedipalps,* lie behind the chelicerae (except in horseshoe crabs)
4. There usually are four pairs of walking legs, some of which may be specialized for sensory and manipulatory functions
5. There are no paired appendages on the abdomen
6. The organs of excretion and osmoregulation are Malpighian tubules and coxal glands

B. Class Merostomata

1. Commonly called *horseshoe crabs,* this class has only five living species
2. All are benthic marine organisms; they crawl over the shallow ocean bottom feeding on small invertebrates and algae
3. Horseshoe crabs have an unsegmented, horseshoe-shaped carapace, which gives the group its common name
    a. Their broad, fused abdomen has a long, spine-like tail *(telson),* which is used to push the body forward in burrowing or for righting the animal when it is overturned
    b. The posterior region of the abdomen contains up to 200 leaf-like gills, collectively called *book gills*
4. Horseshoe crabs congregate to mate during the full moons of spring and early summer
    a. The male clings to the female's abdomen and remains attached until mating is completed; the second pair of appendages (the pedipalps of other species) are modified for grasping and holding

   b. The male releases sperm onto the eggs as the female deposits them in the sand
   c. After the partners detach, the eggs are covered with sand
5. Development is indirect; a *trilobite larva* (which resembles trilobites, extinct members of this subphylum) burrows into the sand and develops into an adult
   a. Growth and development are slow
   b. Sexual maturity is not reached until about 9 years of age
6. The most commonly seen example is the horseshoe crab *Limulus*

C. Class Arachnida

1. This large and successful class has more than 50,000 living species; the 13 orders include spiders (order Araneae), scorpions (order Scorpionida), and ticks and mites (order Acarina)
2. All arachnid species are terrestrial and live in a variety of habitats

D. Order Araneae

1. Spiders are characterized by the lack of obvious external segmentation in the cephalothorax and abdomen (see *Internal Anatomy of a Female Spider*); the abdomen is coupled to the cephalothorax by a slender waist
2. The more than 35,000 described species of spiders are distributed worldwide
3. As predatory carnivores, spiders primarily feed on insects
   a. The chelicerae function as poisonous fangs
      (1) Neurotoxic venom, which is transported in ducts from poison glands into the chelicerae, is injected to subdue the prey

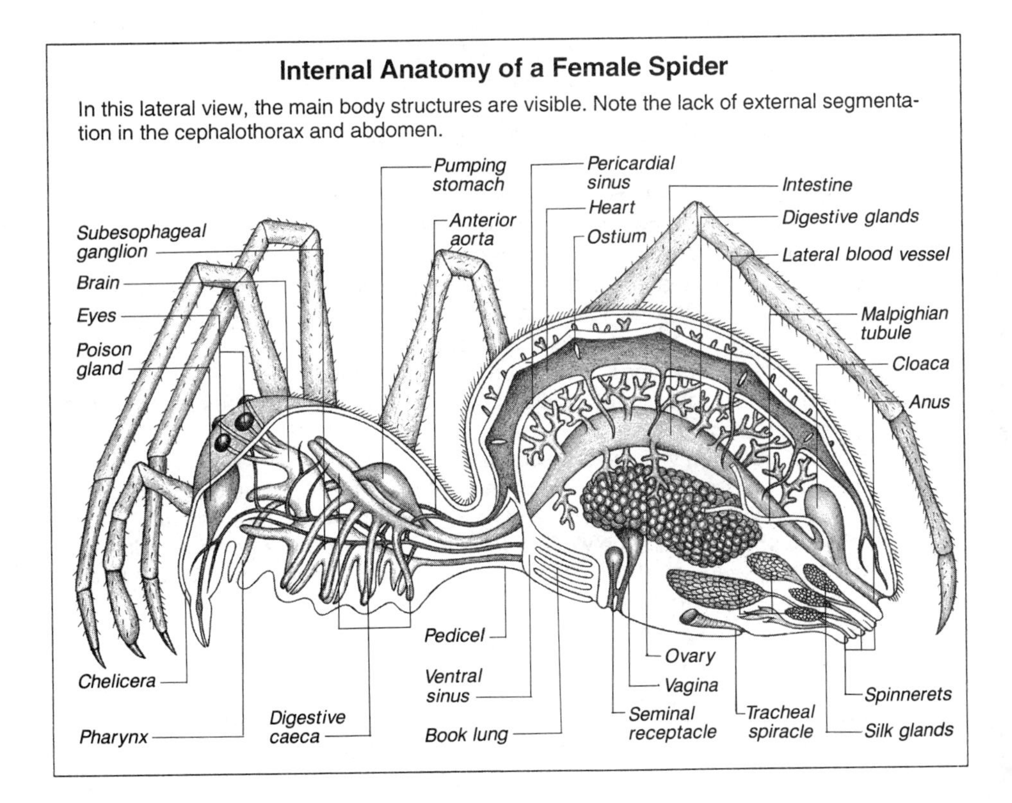

Internal Anatomy of a Female Spider

In this lateral view, the main body structures are visible. Note the lack of external segmentation in the cephalothorax and abdomen.

(2) The venom of some spiders, such as the American black widow *(Latrodectus mactans)* and the brown recluse *(Loxosceles reclusa)*, is powerful enough to cause illness and even death in humans

b. The chelicerae of most spiders have teeth that crush the prey

c. Enzymes flood the prey's body to complete external digestion; this reduces the prey tissues to a liquid, which then is sucked into the stomach

4. Spiders have book lungs, tracheal systems, or both
   a. *Book lungs* consist of many leaf-like air pockets extending into the hemocoel; they open to the outside via spiracles (also called *lung slits*)
   b. The "pages" of the book lungs interdigitate with the hemocoel fluids, providing extensive contact areas for gas exchange
   c. The tracheal system is similar to that found in insects, but smaller and less complex
5. Spiders usually have eight simple eyes, arranged in two rows of four; although vision usually is poor, some spiders locate their prey by sight
6. The body is covered with tactile hairs that sense vibrations, air currents, touch, and some sound frequencies; these receptors are highly developed
7. Chemoreceptors are found in the walking legs and pedipalps; they sense liquid and airborne chemicals (comparable to the human senses of taste and smell)
8. Spiders exhibit complex courtship and mating rituals; reproductive behavior typically includes parental care of the eggs or young
   a. Male spiders have a single pair of tubular testes
      (1) Sperm are released into a sperm duct and carried to the male gonopore
      (2) Sperm are stored and transferred by the pedipalps, which also serve as copulatory organs
   b. Female spiders have a pair of ovaries; the oviducts unite to form a vagina (also called a *uterus*) (see *Internal Anatomy of a Female Spider,* page 105)
      (1) Sperm packets are placed by the male into the female's seminal receptacles or are left on the ground for the female to pick up
      (2) The sperm are stored until the female lays her eggs, which may take place months after copulation
   c. Groups of fertilized eggs are enclosed in silken egg cases (cocoons), where they complete their development; the female may carry the cocoons on her body or attach them in protected locations
   d. Young spiders that hatch are able to feed, but may be in various stages of development, depending on the species; in some cases, parental care continues until the young can fend for themselves
9. Spiders have various types of silk-producing structures
   a. The silk is a fibrous protein with unique properties; the strands are extremely strong and flexible
   b. Silk is secreted in liquid form but hardens on contact with air
   c. It may be used in web construction or as a method of transportation; it also is used to line burrows, form eggs cases, or assist with prey capture
10. Examples of spiders are the wolf spider *(Lycosa aspera)*, the golden garden spider *(Argiope aurantia)*, the black widow *(Latrodectus mactans)*, and the tarantula *(Dugesiella hentzii)*

E. Order Scorpionida

1. Scorpions are the most primitive arachnids and among the earliest terrestrial arthropods
2. All of the approximately 1,200 living species of scorpions are predatory carnivores
    a. They feed primarily on insects, but larger forms also may feed on snakes and lizards
    b. They detect prey with highly sensitive tactile receptors
3. Scorpions are found primarily in warm climates, especially deserts and tropical forests; most species are nocturnal
4. The scorpion cephalothorax is short, but the abdomen is elongated and segmented
    a. Appendages are attached to the cephalothorax
    b. The large, claw-like pedipalps are used to grasp prey
    c. The short chelicerae are adapted for tearing and grinding food
    d. The abdomen terminates in a sharp, venomous sting
        (1) Scorpion venom is a powerful neurotoxin that can quickly paralyze and kill the prey
        (2) The venom of some species can kill larger animals, including humans
5. Gas exchange occurs via book lungs
6. Scorpions exhibit complex courtship behavior patterns
    a. Mating may occur in a series of ritualized "dance" steps, which culminate in the release of a sperm packet by the male
    b. The female positions herself over the sperm packet, the packet bursts, and sperm enter the female gonopore
    c. Scorpions give birth to live young that are carried on the mother's back until the first molt
7. An example of a scorpion is the striped scorpion *(Centruroides vittatus)*

F. Order Acarina

1. Acarinids (ticks and mites) are the largest arachnid group; more than 30,000 species have been described, and evidence suggests that many more await discovery
2. Acarinids are distributed worldwide
    a. Their tiny, compact bodies enable them to adapt to a variety of microhabitats
    b. Free-living forms may be terrestrial or aquatic; most prey on small invertebrates
    c. Many species are symbiotic or parasitic; the hosts often are other arthropods but may be vertebrates, including humans
3. Acarinids are characterized by the complete fusion of the cephalothorax and abdomen, with no externally visible segmentation
4. Respiration usually is by book lungs, but may occur through the skin
5. Ticks, the largest acarinids, are blood-sucking parasites of reptiles, birds, and mammals
    a. Their chelicerae are adapted for piercing the skin
    b. Ticks may remain attached to the host for several weeks, feeding on the host's body fluids
    c. Several ticks are *vectors* (intermediate hosts) for serious human diseases, such as Rocky Mountain spotted fever and Lyme disease
6. Mites are very small (usually less than 1 millimeter long)

a. Free-living forms may be herbivores or predators; they are found in terrestrial, marine, and freshwater habitats
b. Parasitic species infect invertebrates, vertebrates, and even plants
c. Many mites are economically important pests of crops, but others have been used effectively in *biologic pest control* (an environmentally sound method of removing crop pests without the use of chemicals)
d. A number of mites are important disease vectors of plants, especially viral crop infestations
e. Mites also cause tumors and skin irritations in animals, including humans

7. Examples of acarinids are the wood tick *(Dermacentor variabilis)*, the red velvet mite (*Trombidium* species), and *Dermacentor andersoni* (the vector of Rocky Mountain spotted fever)

VI. Subphylum Crustacea

A. General information

1. Crustaceans are an abundant group with more than 30,000 described species; experts believe that many times that number have yet to be discovered
2. Members of this subphylum primarily are aquatic and are found in marine and freshwater habitats; a few species, including the pill bugs, are terrestrial
3. Crustaceans are characterized by two pairs of antennae on the head
   a. Feeding appendages include a pair of heavy *mandibles* and two small, manipulatory pairs of *maxillae*
   b. Posterior abdominal appendages may be modified for copulation (in males) or for brooding eggs or young (in females)
   c. The first pair of walking legs may bear strong claws, which are used for defense
4. The crustacean body may have 16 to 60 somites
   a. More advanced species have fewer somites, which are grouped into tagmata; the principal tagmata are the head, thorax, and abdomen
   b. A carapace is usually present
      (1) In some species, the carapace covers a small area of the dorsal surface
      (2) In others, the carapace is greatly enlarged; it may extend over the head, dorsal, and lateral body surfaces and may fuse with thoracic segments to form a cephalothorax
5. Gas exchange occurs by gills
   a. The gills usually are attached to the appendages or may be enclosed in branchial chambers between the carapace and the body wall
   b. Vibrating or beating appendages draw a current of water over the gills when the crustacean is not moving
6. Antennal glands (also called *green glands*) function as the organs of excretion and osmoregulation
7. Crustaceans have well-developed nervous systems and sense organs
   a. The cerebral ganglia lead to a ventral nerve cord with segmental ganglia (see *Internal Structure of a Male Crayfish,* page 104)
   b. Some crustaceans have simple ocelli, but most have compound eyes (similar to those of insects)

c. The body surface is covered with sensory receptors, including tactile receptors, chemoreceptors, statocysts (which also function as rheoreceptors), and possibly thermoreceptors

8. Most crustaceans are dioecious and have a variety of reproductive specializations
   a. Development usually is indirect with a *nauplius* larval stage
   b. Other larval stages may follow the nauplius stage
9. Classification of crustaceans, although still controversial, is based primarily on the number of somites in the thorax and abdomen, the style of the appendages, and the size and shape of the carapace

B. Class Ostracoda

1. Ostracods are small animals, ranging from 0.1 to 32 millimeters in length
2. The 2,000 living species are numerous and diverse in habitat
3. Ostracods are filter feeders, herbivores, or deposit feeders
4. The ostracod body is enclosed in a bivalved carapace that externally resembles a tiny clam; ostracods have few appendages compared with most other crustaceans
5. Examples of ostracods are *Cypridina, Gigantocypris* (a huge ostracod over 3 centimeters long), and *Cypris*

C. Class Copepoda

1. Copepods are a numerous and important group of crustaceans
2. Of the 9,000 described species, most are less than 10 millimeters long, but some grow to 1.5 centimeters
3. Members of this class may be free-living, symbiotic, or parasitic
   a. Free-living forms are detritus feeders, herbivores, or omnivores that feed on diatoms, bacteria, and protozoans
   b. Parasitic forms are endoparasites or ectoparasites of fish and invertebrates
4. Copepods have four or five pairs of walking (or swimming) appendages attached to the thoracic region; there are no legs on the abdomen
5. They are extremely abundant and very important in aquatic food chains, where they often are the dominant primary consumers
6. Examples of copepods are *Cyclops, Clavella,* and *Harpacticus*

D. Class Cirripedia

1. Members of the class Cirripedia, also called *barnacles,* once were classified with the molluscs, because their body is enclosed in a shell of fused calcareous plates
2. Inside the shell, however, is a highly modified, but recognizable, arthropod body (see *Internal and External View of the Goose-Necked Barnacle Lepas,* page 110)
   a. The head is reduced and the abdomen is absent
   b. The segmented thorax bears long limbs with multiple joints, called *cirri,* that are adapted for filter feeding
3. In most species, the planktonic larva is the only mobile stage of the life cycle; adults attach to hard substrates, including floating objects and the bodies of marine animals (such as whales or turtles)
4. Most species are hermaphroditic, although some groups are dioecious
5. Examples of barnacles are the goose-necked barnacle *(Lepas fascicularis)* and acorn barnacles (such as *Cthalamus* species and *Balanus* species)

Internal and External View of the Goose-Necked Barnacle *Lepas*

In the external view (A), note the calcareous shell. Internally (B), note the presence of jointed appendages (cirri), adhesive gland, and both male and female reproductive organs.

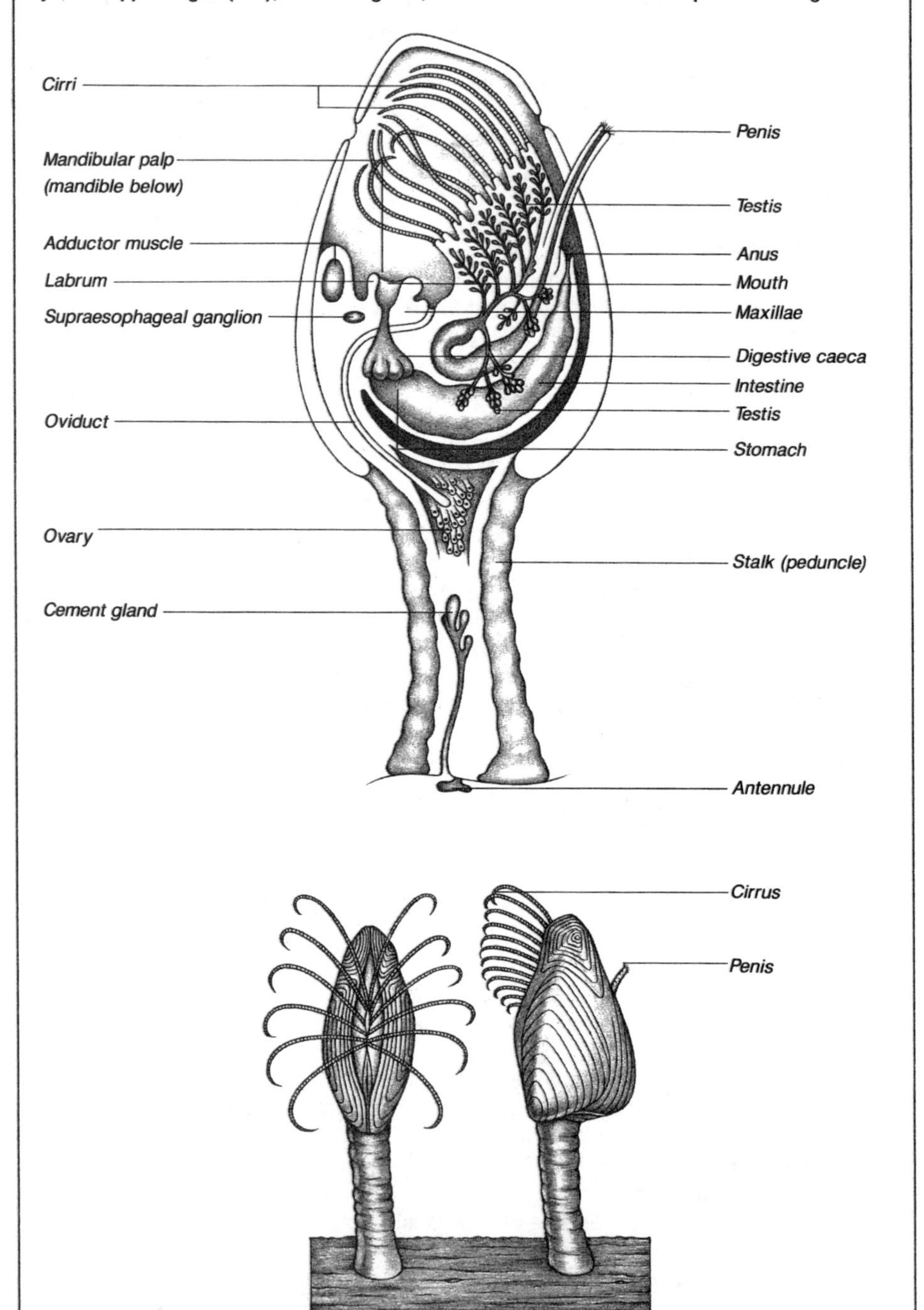

E. Class Malacostraca

1. With 13 orders, class Malacostraca is the largest crustacean class
2. Members of this class have a eight-segment thorax and a six-segment abdomen; each of these segments has a pair of appendages
3. Class Malacostraca contains many economically and ecologically important species, including the isopods, amphipods, euphausids, and decapods
4. The *isopods* (order Isopoda) are dorsoventrally flattened and have abdominal gills
    a. Usually about 1 centimeter in length, isopods do not have a carapace
    b. Isopods commonly are found on beaches and rocky shores
    c. Terrestrial species are called *pill bugs, wood lice,* or *sow bugs;* they are characterized by the ability to roll up into a ball when threatened
    d. Isopods are omnivores and usually scavengers
    e. Respiration is by gills on the abdominal appendages
    f. Development is direct; females brood the young in special chambers
5. The *amphipods* (order Amphipoda) are small and laterally flattened
    a. The carapace is absent, and the gills are located in the thorax region
    b. Most amphipods are found in benthic marine habitats, but some are freshwater and terrestrial species
    c. Many burrow in the substrate or live in tubes; others, such as the skeleton shrimp, live symbiotically on the bodies of sessile invertebrates
6. The *euphausids* (order Euphausiacea) are shrimp-like crustaceans about 4 to 15 centimeters in length
    a. Commonly known as *krill,* the euphausids are a vital component of oceanic food chains
    b. These pelagic animals live in vast swarms (up to half a mile in diameter), especially in polar seas
    c. They are eaten by baleen whales, squid, marine birds, and many fish
    d. Krill primarily are filter feeders; some are predators or detritus feeders
7. The *decapods* (order Decapoda) include many important food species
    a. Order Decapoda is the largest order of crustaceans, containing more than 10,000 species, including the lobsters, crayfish, shrimp, and crabs
    b. Decapods have five pairs of walking legs, which give the order its name (decapoda means "ten feet"); the first pair of legs often is modified into large claws called *chelipeds,* which are used for predation and defense (see *Internal Structure of a Male Crayfish,* page 104)
    c. Three pairs of small appendages called *maxillipeds* function in food handling
    d. Decapods have a well-developed carapace; the gills are enclosed in a branchial chamber formed by the carapace and ventilated by modified appendages
    e. In crabs, the abdomen is reduced and folded beneath the cephalothorax, producing a shortened body
    f. Compound eyes are located on stalks at the anterior end of the cephalothorax
    g. Decapods are found in diverse habitats and employ all types of feeding strategies; most are either predators or scavengers
    h. The digestive tract has complex regional specializations
        (1) The foregut is divided into an anterior *gastric stomach,* which contains a tooth-like gastric mill used for grinding food particles, and a posterior *pyloric stomach*
        (2) Large food items can be ingested and stored in the enlarged gastric stomach for later processing and digestion

i. Decapods are dioecious and have a single pair of gonads
(1) During copulation, sperm packets are transferred to the female gonopore
(2) Fertilized eggs are brooded on the ventral side of the female abdomen
(3) Some decapods hatch as nauplius larvae, but most pass through that stage within the egg, hatching at a later planktonic larval stage called a *zoea*
(4) Most freshwater species, such as the crayfish, exhibit direct development; eggs are brooded until development is complete

8. Examples of decapods are the fiddler crab *(Uca minax),* the blue crab *(Callinectes sapidus),* and the Northern lobster *(Homarus americanus)*

VII. Subphylum Uniramia

A. General information

1. Subphylum Uniramia is the largest arthropod class, with more than 1 million described species of insects alone; experts suggest that 20 to 50 million additional species remain undiscovered
2. Most uniramians live in terrestrial habitats; aquatic forms are secondarily adapted for this life-style
3. Uniramians include the centipedes (class Chilopoda), the millipedes (class Diplopoda), the insects (class Insecta), and others
   a. They are characterized by unbranched appendages, which give the subphylum its name
   b. They have only one pair of antennae, and the carapace is absent
   c. As with crustaceans, the feeding appendages include a pair of heavy mandibles and two small, manipulatory pairs of maxillae
   d. The exoskeleton is highly sclerotized, strengthening the body and appendages
   e. Respiration is by tracheae and spiracles; aquatic larval forms may have gills
   f. The organs of excretion and osmoregulation are Malpighian tubules; specialized rectal glands, if present, increase reabsorption of water, ions, and other nutrients
4. Uniramians are dioecious; some species (especially among the insects) may have complex reproductive behaviors

B. Class Chilopoda

1. Centipedes average about 1 to 2 centimeters in length, but some of the 2,500 described species grow to 25 centimeters
2. Centipedes are found primarily in moist habitats of temperate and tropical regions
3. All centipedes are predatory carnivores, showing several specializations for rapid locomotion and prey capture
   a. They feed on worms, snails, and various arthropods
   b. Large species may prey on small lizards or frogs
4. The centipede body is dorsoventrally flattened
   a. The body may have up to 173 trunk segments
   b. Except for the most anterior and posterior segments, each segment has a pair of walking legs

c. The legs on the first trunk segment (maxillipeds) are modified into large poisonous claws that are used to stab prey and inject a paralyzing venom
d. The cuticle is not calcified and has no waxy layer
e. Only one pair of Malpighian tubules is present

5. Centipedes are dioecious and usually have some form of mating behavior
   a. Mating rituals culminate in the female picking up the sperm packet and inserting it into her gonopore
   b. Eggs are laid in rotting vegetation or on the ground
   c. Most species guard the eggs until hatching occurs
   d. Development is direct, with no larval stages

C. Class Diplopoda

1. Millipedes average about 1 centimeter in length but may grow to 30 centimeters; approximately 10,000 species have been identified
2. Most species are herbivores and feed primarily on dead and decaying vegetation; a few species are predatory carnivores and resemble centipedes in their diet
3. Millipedes are ecologically significant in the mixing and recycling of leaf litter
4. Millipede bodies are cylindrical in cross section, with up to 100 segments
   a. The smooth, rounded body and many legs are well suited for pushing and burrowing through foliage and forest floor debris
   b. The cuticle is highly sclerotized and often strengthened by calcareous deposits
   c. The head is separated from the trunk by an enlarged, limbless collar (the first thoracic segment)
   d. The mandibles are adapted for tearing and chewing vegetation
   e. Each of the next three body segments has one pair of walking legs
   f. The rest of the body is composed of fused *diplosegments* (pairs of segments fused together), which have two pairs of walking legs each
   g. Metameric organs and body structures (such as ganglia) are also doubled in each diplosegment
   h. Two pairs of Malpighian tubules are present
5. Millipedes are dioecious and have complex mating behaviors
   a. Copulating pairs coil together and may remain attached for several days
   b. The sperm packet is placed in contact with the female gonopore
   c. Enzymes dissolve the sperm packet casing, and the sperm migrate into the seminal receptacles of the female
   d. Eggs are fertilized as they are laid
      (1) Some species construct nests of soil and decayed vegetation, cemented together with feces
      (2) Some nests are entirely of fecal origin
   e. Development is direct, but juveniles have a reduced number of segments and appendages, which are added as the individual grows

D. Class Insecta

1. Insects are the most numerous and successful of the arthropods; they are distributed globally and found in most known habitats
2. Many symbiotic insect species have been described; however, most insects are ectoparasitic, and many are vectors for serious diseases of humans and domestic livestock

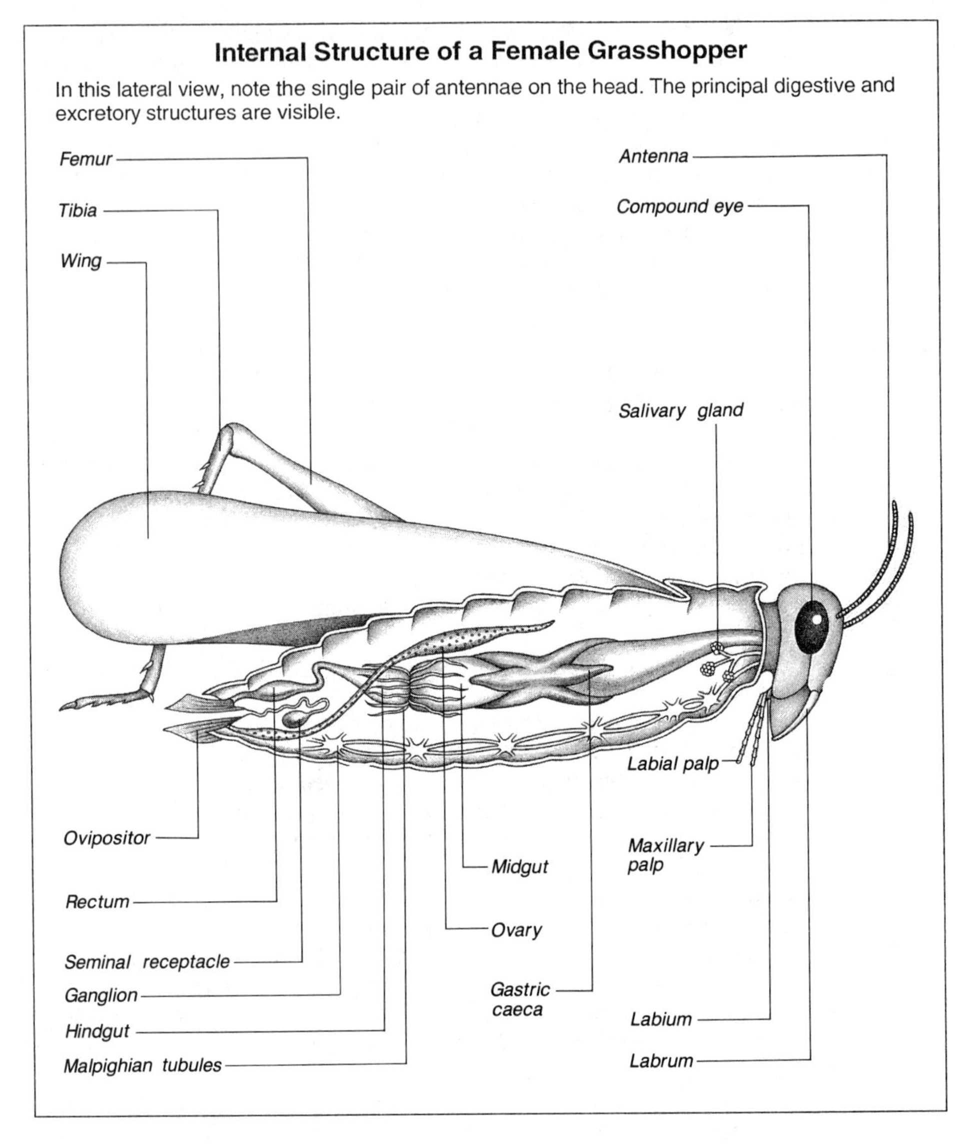

3. Insect tagmata consist of the head, thorax, and abdomen (see *Internal Structure of a Female Grasshopper*)
   a. The head bears a large pair of compound eyes and a single pair of antennae
      (1) The feeding appendages are mandibles and two pairs of maxillae
      (2) The second pair of maxillae are fused into a flap-like lower lip, called a *labium*
      (3) The upper lip, called the *labrum,* is an extension of the head
   b. The thorax has three segments, each with a pair of walking legs
      (1) The walking legs often are modified for specific life-styles
      (2) Modifications include sticky pads, claws, paddles for swimming, and long, athletic jumping legs

c. Wings, when present, are attached to the last two thoracic segments
d. The abdomen has 11 segments, some of which may be fused; the abdominal segments do not have legs
4. Insects have a waxy cuticle, which prevents desiccation and repels water
5. Mouthparts are modified for a variety of feeding modes; the three basic feeding methods are biting-chewing, sucking, and sponging
a. *Biting-chewing insects* may be herbivores, carnivores, or scavengers
(1) Common examples are locusts, crickets, and grasshoppers
(2) The maxillae and labium have manipulatory palps, which hold the food in position while the mandibles bite or tear the food and pass pieces to the mouth (see *Internal Structure of a Female Grasshopper*)
b. *Sucking insects* feed on plant nectars or saps, or may be parasitic feeders on the blood and body fluids of a host
(1) The mouthparts are small and form a long tube
(2) In parasitic forms, the salivary glands often secrete digestive enzymes into the wound to predigest tissues; blood feeders secrete anticoagulants
c. In *sponge-feeding insects,* such as the house fly and fruit fly, the labium forms an expanded porous surface that absorbs nutrients
(1) Saliva is secreted onto the food to dissolve it
(2) The liquified nutrients are "sponged up" by the labium and transported to the mouth
(3) Biting spongers, such as the horse fly, make a wound with the mandibles and then sponge up body fluids
6. The insect digestive system is complex and regionally specialized
a. The *foregut* has several parts
(1) The mouth and salivary glands are associated with food-handling appendages
(2) The muscular pharynx may be modified into a pharyngeal pump in sucking forms
(3) The esophagus leads to a crop for food storage; species that feed infrequently have highly elastic crop walls
(4) Biting and chewing species have a food-grinding organ (gizzard)
b. The *midgut* contains the stomach and several gastric ceca; it is the primary site of nutrient absorption
c. The *hindgut* functions in osmoregulation (water reabsorption) and, to a limited extent, in absorption of nutrients
d. Many herbivorous species have symbiotic bacteria and protozoans living in the midgut or hindgut regions; these symbiotes are involved in cellulose digestion
7. The heart extends into several abdominal segments; the number of ostia varies
a. Accessory pumping organs are located at the base of wings and long appendages (such as the hind legs of grasshoppers)
b. Multiple pumps help to overcome the sluggish nature of the open circulatory system and facilitate movement in active forms
8. An extensive tracheal system functions in gas exchange
a. Spiracles are the external openings of the tracheal tubes
b. Aquatic larvae and adults have tracheal gills
9. Malpighian tubules are the organs of excretion and osmoregulation
10. Sensory receptors are complex and well developed

a. Insects have both compound and simple eyes
    (1) Research suggests that increased visual acuity is correlated with an increased number of ommatidia
    (2) Predatory insects, such as dragonflies, have upward of 10,000 ommatidia in each eye
b. *Proprioceptors* (organs of balance and equilibrium) are well developed, especially in jumping and flying forms
c. Sound plays an important role in the lives of many insects; consequently, sound receptors are complex and highly developed
d. Chemoreceptors are distributed over the body surface, especially on the antennae, mouthparts, and legs
e. Others sensory receptors include tactile receptors, temperature and humidity sensors, and georeceptors

11. Insects are dioecious and may display complex reproductive behavior
    a. In many species, females secrete attractant chemicals *(pheromones)* that can be sensed over great distances (up to several miles) by males
    b. Fertilization usually is internal; the male inserts his penis into the female reproductive tract and releases seminal fluid
    c. Sperm are stored in seminal receptacles, and eggs are fertilized when laid
    d. Most species mate infrequently (as little as once in a lifetime); sufficient sperm are stored for several successive egg-laying sessions
    e. The number of eggs varies with the species, from as few as one up to several thousand at a time
12. More than 80% of all insects undergo *complete metamorphosis*
    a. Larvae differ extensively from the adults, as seen in caterpillars versus butterflies
    b. When sufficient size and development is achieved, larvae *pupate*
        (1) The larva forms a case or cocoon around its body (the *pupa*) and remains dormant within the pupa while the body is transformed
        (2) Energy reserves stored during the larval stage are used to reconstruct the body into its adult form
        (3) The insect emerges from the cocoon as a sexually reproducing adult
13. The remaining 20% of insects undergo *incomplete metamorphosis*
    a. The young (usually called *nymphs*) resemble the adults but lack sexual structures and functional wings
    b. Nymphs grow gradually, and through successive molts, develop into sexually reproducing adults
    c. There is no dormant or pupa stage
14. Insects are capable of complicated social behaviors
    a. Communities may have up to 70,000 individuals (as in a honeybee hive)
    b. Specialized individuals have specific functions in the community (such as worker, drone, or queen)
    c. Complex community structures are formed by termites, ants, and bees

Study Activities

1. List seven basic characteristics of the arthropods.
2. Explain current hypotheses concerning the evolutionary ancestry of arthropods.
3. Prepare a table or chart that compares and contrasts the ways various arthropod groups carry out nutrition, gas exchange, and excretion.
4. Describe methods of reproduction in arthropods.
5. Discuss the basic characteristics of each of the arthropod subphyla and give examples of each.
6. Compare and contrast the classes and orders of subphylum Chelicerata.
7. Compare and contrast the classes of subphylum Crustacea.
8. Compare and contrast the classes of subphylum Uniramia.

12

Lophophorate Phyla: Bryozoa, Phoronida, and Brachiopoda

Objectives

After studying this chapter, the reader should be able to:
- Describe the basic characteristics of the lophophorate phyla.
- Explain current hypotheses concerning the evolutionary ancestry of the lophophorates.
- Describe how bryozoans perform their basic life functions.
- Compare and contrast the basic characteristics of the bryozoans with those of other lophophorate phyla.

I. Basic Characteristics

A. General information

1. The three lophophorate phyla are superficially dissimilar
    a. The bryozoans (phylum Bryozoa or Ectoprocta) are tiny, colonial animals that form encrusting colonies on hard substrates
    b. The phoronids (phylum Phoronida) are benthic marine animals that resemble worms and live in secreted tubes
    c. The brachiopods (phylum Brachiopoda) have bivalved shells similar to those of molluscs
2. Lophophorates share many common characteristics
    a. They are deuterostome eucoelomates that have a three-part body plan
        (1) Each body region contains an individual coelomic compartment
        (2) The coelomic cavities often are paired
    b. They have simple reproductive and digestive systems
    c. Most secrete protective outer casings, such as tubes, shells, or exoskeletons
3. The group is characterized by a special feeding structure called a *lophophore* (a cluster of ciliated tentacles), which is adapted for filter feeding
    a. The hollow tentacles, which surround the mouth but not the anus, contain an extension of the coelom
    b. When the lophophore is extended, it forms a funnel that directs food into the centrally located mouth
    c. The thin-walled tentacles also may function in gas exchange
4. Lophophorates have a U-shaped gut, which is an adaptation for a sessile, enclosed life-style (within a tube or shell) (see *Representative Lophophorate Phyla*, page 123)

a. The curved gut brings the anus to the exterior of the enclosure, thereby preventing fouling
b. Ciliary currents from the lophophore carry wastes away

B. Ecologic relationships

1. Most lophophorates are exclusively marine, but a few bryozoans are found in freshwater habitats
2. All lophophorates have a benthic life-style, living in secreted tubes, shells, or exoskeletons
3. All members of the lophophorate phyla are filter feeders

C. Evolutionary relationships

1. The origin of the lophophorates is not completely clear
2. They may have arisen with the deuterostome line; however, the origin of deuterostomes also is unclear
3. The ancestral lophophorate may have been a coelomate burrowing form similar to modern phoronids; it was probably a filter feeder
4. Other lophophorate phyla diverged from this basic body plan

II. Basic Characteristics of Phylum Bryozoa

A. General information

1. Bryozoans are tiny animals enclosed within a protective exoskeleton; most are less than 0.5 millimeters long
2. Bryozoans form colonies of polymorphic individuals; for this reason, they sometimes are referred to as *moss animals*
3. They have no specialized structures for circulation, respiration, or excretion
4. Bryozoans are hermaphroditic, and development includes a planktonic larval stage

B. Ecologic relationships

1. Bryozoans are primarily marine, but a few species are found in freshwater habitats
2. They attach to a variety of substrates, including rocks, shells, vegetation, and even the hulls of ships; bryozoan colonies recently have been discovered attached to ice flows in the Antarctic
3. Bryozoan colonies can be quite large, covering rocks and even creating miniature reef systems
4. About 4,500 living species of bryozoans have been described, and they are well represented in the fossil record

III. Bryozoan Form and Function

A. General information

1. Bryozoans have adaptations for a sessile, filter-feeding life-style
2. They superficially are similar to hydrozoans (see Chapter 7, Radiate Phyla: Cnidaria and Ctenophora), but differences are apparent under magnification
   a. Bryozoans have ciliated lophophore tentacles

b. Bryozoans are encased within a chitinous or gelatinous exoskeleton; a layer of calcium carbonate usually is incorporated in the exoskeleton
3. A bryozoan colony is produced by asexual budding from a single zooid (individual animal)

B. Body types of bryozoans
1. Individual zooids are shaped like a box or cylinder (see *Representative Lophophorate Phyla,* page 123)
   a. The lophophore extends through an opening in the exoskeleton
   b. This opening often is covered with a flap-like lid, called an *operculum*
2. Colony members are attached together, and different species exhibit different colonial growth patterns
   a. In some groups, the zooids grow separately from horizontal runners, called *stolons*
   b. In other groups, the zooid exoskeletons are attached directly to each other
   c. In both cases, all zooids in a colony are interconnected
   d. Pores in the exoskeleton allow diffusion among colony members, and in some species, there are tissue connections between individual zooids
3. Various types of zooids are found in bryozoan colonies
   a. *Autozooids* are responsible for feeding and digestion; no other zooid forms are capable of feeding
   b. *Kenozooids* are simplified individuals with special attachment structures; they secure the colony together or to the substrate
   c. *Avicularia* are specialized for colony defense; the operculum is modified into a hinged beak, which is used to deflect small organisms and to pick debris from the colony surface

C. Reproduction
1. New colony members are produced by asexual budding; new colonies are produced by sexual reproduction
2. Most bryozoans are hermaphroditic
   a. The gonads release gametes into the coelom
   b. Sperm exit through pores in the end of specialized tentacles
   c. Eggs exit through a pore at the base of two tentacles
3. Some species shed eggs into the water, but most brood the eggs internally
   a. Because eggs are brooded internally, they tend to be few in number
   b. Some bryozoans have specialized zooids, called *gonozooids,* that brood embryos
4. Development usually is indirect, with a planktonic larval stage that is similar to a trochophore larva; the planktonic larva settles onto a substrate, attaches to it, and begins a new bryozoan colony

D. Circulation, gas exchange, excretion, and osmoregulation
1. Bryozoans have no special organs for circulation, gas exchange, or excretion
2. They are small enough that these functions can be accomplished by diffusion and movement of the coelomic fluid
3. Nitrogenous wastes are excreted as ammonia
4. Protonephridia are the organs of osmoregulation

E. Nervous system and sense organs

1. The bryozoan nervous system and sense organs are reduced
2. A single ganglion and nerve ring surround the anterior end of the gut
3. The only sensory structures are tactile receptors

IV. Basic Characteristics of Phylum Phoronida

A. General information

1. Phoronids are marine animals with simplified, worm-like bodies; they range from 5 to 25 centimeters in length
2. Phoronids live in secreted chitinous tubes, which may be anchored to a hard substrate or buried in soft sediments

B. Ecologic relationships

1. Phoronids are found in intertidal mud flats and up to a depth of 400 meters
2. This small group has about 10 identified species
3. Phoronids are sessile filter feeders

V. Phoronid Form and Function

A. General information

1. The body is simplified with little regional specialization
    a. The lophophore tentacles are arranged in ridges (see *Representative Lophophorate Phyla*, page 123)
    b. The end opposite the lophophore is enlarged; it contains the stomach and helps to anchor the animal within the tube
2. Owing to an unusual developmental pattern, anterior and posterior regions are not easily distinguished; body parts are located at the "lophophore end" or the "stomach end"
3. The coelom is divided into three sections; coelomic fluid functions as a hydrostatic skeleton
4. Additional support for the body is obtained from the walls of the secreted tube

B. Reproduction

1. Phoronids are capable of both asexual and sexual reproduction
2. Asexual reproduction occurs by budding or transverse fission; phoronids also can regenerate lost or damaged body parts
3. Some phoronids are dioecious, others are simultaneous hermaphrodites (function as both male and female at the same time in their life cycle)
    a. Gametes are released into the coelomic cavity and carried out through the metanephridia
    b. In some species, internal fertilization occurs in the coelomic cavity
4. Development is indirect and includes a ciliated *actinotroch* larval stage; the larva settles onto a substrate and develops into a tube-dwelling adult

VI. Basic Characteristics of Phylum Brachiopoda

A. General information

1. Brachiopods, commonly called *lamp shells,* are remnants of an extremely successful group of organisms
    a. There are about 335 living species of brachiopods
    b. During the Paleozoic and Mesozoic eras, there were more than 12,000 living species of brachiopods
    c. Some modern species are virtually identical to their fossil relatives
2. Brachiopods externally resemble bivalved molluscs, because they have a hinged, bivalved shell
3. Brachiopods range from 1 millimeter to more than 9 centimeters in length

B. Ecologic relationships

1. Brachiopods are found only in marine, benthic habitats
2. Brachiopods orient themselves on the substrate ventral side up
3. They are nondiscriminatory filter feeders, but phytoplankton constitute a large percentage of the diet

VII. Brachiopod Form and Function

A. General information

1. The brachiopod body is enclosed in a bivalved, hinged shell
2. Brachiopods differ from molluscs in the structure and orientation of the shell valves
    a. The brachiopod shell has dorsal and ventral valves; the mollusc shell has right and left valves
    b. Brachiopod shells usually are unequal in size
3. Brachiopods resemble molluscs in that the shell is secreted by a mantle
    a. The mantle lines the inside of the shell
    b. The lophophore is located inside the mantle cavity
    c. The mantle is formed from folds in the body wall; extensions of the coelom run through the mantle tissue
4. The coelom is divided into three compartments, but the partitioning may be obscured by secondary structural modifications in body form
5. Most brachiopods are attached to the substrate by a fleshy stalk, called a *pedicle* (see *Representative Lophophorate Phyla*); species lacking a pedicle attach directly to the substrate

B. Reproduction

1. Brachiopods are dioecious
    a. Gonads are transient and develop from the peritoneum
    b. Fertilization primarily is external
2. Development is indirect, with a free-swimming larval stage
    a. The larva has not been named officially but is tentatively referred to as a *lobate* larva
    b. After 1 to 2 days, the larva settles to the substrate and develops into an adult

Representative Lophophorate Phyla

Note the size and location of the lophophore, which is characteristic of all lophophorate phyla. The illustrations below also show the body structures for a phoronid, a bryozoan, and a brachiopod. The phoronid digestive system, with its U-shaped gut, is typical of that found in many tube-dwelling animals. In this lateral view of a bryozoan, note the U-shaped gut, the ciliated lophophore, and the exoskeletal pores that connect colony members. As shown here, a brachiopod's ventral valve typically is larger than the dorsal valve.

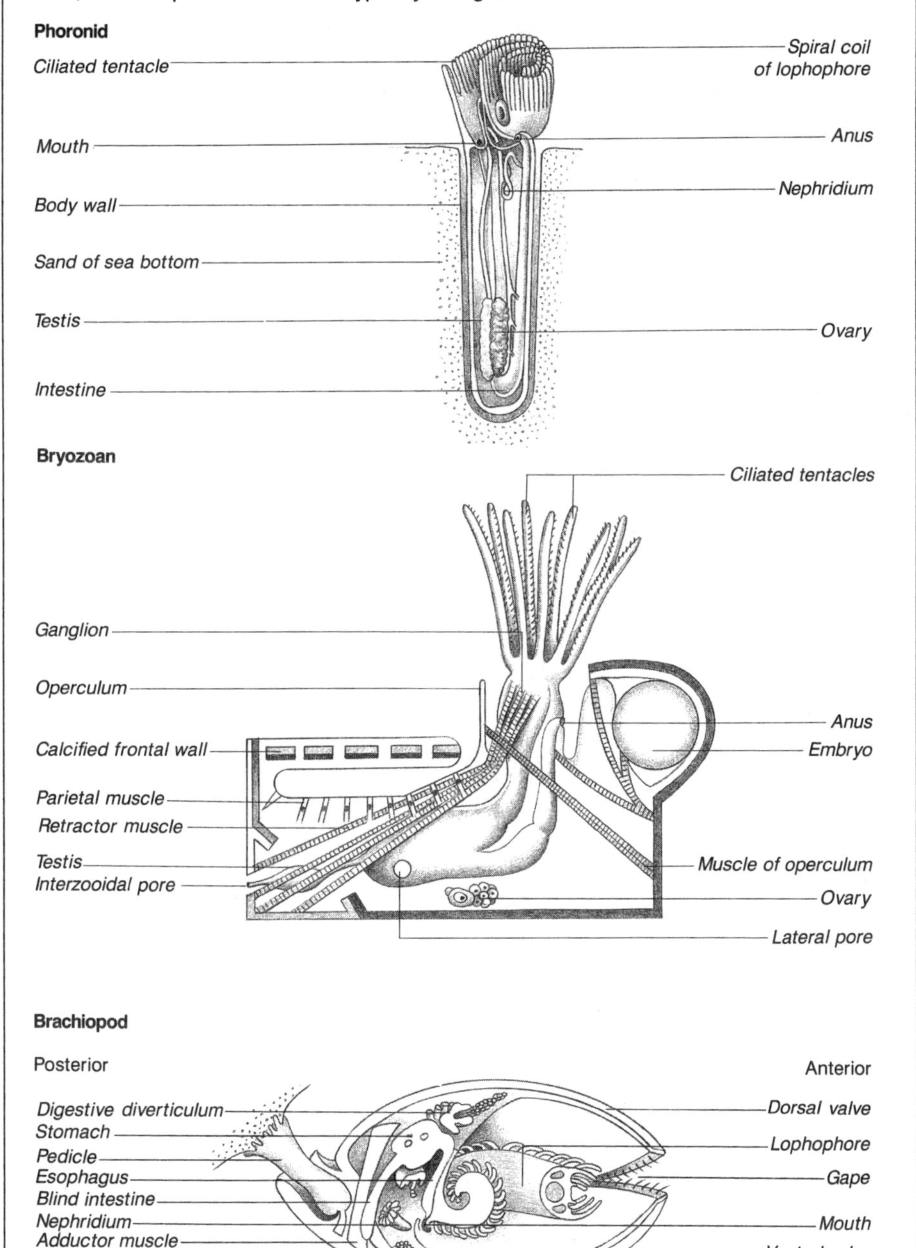

Study Activities

1. List four basic characteristics of the lophophorate phyla.
2. Describe how bryozoans carry out nutrition, gas exchange, and excretion.
3. Describe the methods of reproduction in bryozoans.
4. Compare and contrast the basic characteristics and life-style of bryozoans with those of the other lophophorate phyla.
5. Create a chart or table that compares and contrasts the basic characteristics of bryozoans with colonial hydrozoans.
6. Create a chart or table that compares and contrasts the basic characteristics of brachiopods with bivalve molluscs.

13

Phylum Echinodermata

Objectives

After studying this chapter, the reader should be able to:
• Describe the basic characteristics of echinoderms.
• Explain current hypotheses concerning the evolutionary ancestry of echinoderms.
• Describe how echinoderms perform their basic life functions.
• Identify and characterize the classes of phylum Echinodermata.

I. Basic Characteristics

A. General information

1. Echinoderms are deuterostome eucoelomates that range in size from less than 1 centimeter to more than 1 meter in diameter; sea cucumbers can grow to 2 meters in length
2. The phylum is characterized by a spiny skin and *pentamerous* (five-part) radial symmetry in adults
3. The endoskeleton of calcareous plates is embedded in the body wall
4. The *water vascular system,* found only in echinoderms, is a system of water-filled canals that performs a variety of physiologic functions
5. Typical echinoderms are sea stars, sand dollars, sea urchins, and sea cucumbers

B. Ecologic relationships

1. Approximately 7,000 species of living echinoderms have been identified
2. All echinoderms are marine animals that live in benthic habitats, from the shallow intertidal zone to the deep sea; some forms burrow into the substrate
3. Most species are predatory carnivores, but others are filter feeders, deposit feeders, scavengers, or browsers
4. Echinoderms may be numerous in some habitats; in deep-sea environments, they may comprise more than 90% of the benthic inhabitants

C. Evolutionary relationships

1. Scientists have differing opinions concerning the origin of the echinoderms
2. Echinoderms may have arisen from a bilaterally symmetrical ancestor, since radial symmetry in adult echinoderms is secondarily derived (larval echinoderms have bilateral symmetry)
3. Several branches of radially symmetrical echinoderms may have arisen from a free-moving ancestor

4. A more commonly accepted view is that echinoderm ancestors developed radial symmetry as an adaptation for a sessile life-style; the water vascular system arose as an adaptation for sessile filter feeding
   a. Ancestral species were sessile burrowers or tube dwellers
   b. Free-moving echinoderms arose from these sessile ancestors
      (1) This evolutionary development is supported by the primitive echinoderms, such as the crinoids (sea lilies), which are sessile animals with radial symmetry
      (2) Other primitive groups also have a sessile, stalked phase in their life cycles

II. Echinoderm Form and Function

A. General information

1. Echinoderms are unsegmented animals with pentamerous radial symmetry
   a. Body parts are arranged in multiples of five
   b. The body usually is divided into *oral* and *aboral* (the side opposite the mouth) surfaces
2. The canals and reservoirs of the water vascular system extend throughout the entire body and connect to exterior tube feet

B. Internal and external anatomy

1. The echinoderm body wall has several layers
   a. The *epidermis* covers the external body surface
   b. The *dermis* is derived from mesoderm and secretes the skeletal elements
   c. A *muscular layer* lies beneath the dermis
   d. The *peritoneum* lines the body cavity
2. The *ossicles* (skeletal elements) are composed primarily of calcium carbonate, with some magnesium carbonate
   a. They may be *simple* (single) or *compound* (fused into plates)
   b. Spines and bumps *(tubercles)* are outward extensions of the ossicles; movement of spines is under voluntary control
   c. The ossicles also produce pincer-like *pedicellariae,* which are specialized defensive structures
      (1) Pedicellariae remove debris from the skin surface, protect the delicate papulae (skin gills), and may function in prey capture
      (2) Pedicellariae respond by reflex to tactile and chemical stimuli; this response is independent of the main nervous system
      (3) The jaws of the pedicellariae pincer are opened and closed by muscular action
3. *Papulae* (skin gills) are distributed over the external surface, between the ossicles
4. A large, fluid-filled coelom is present
5. The water vascular system is an internal system of canals and reservoirs (see *Water Vascular System of a Sea Star*)
   a. Water enters the system through a specialized skeletal plate, called a *madreporite* or *sieve plate*
   b. The madreporite connects to a *stone canal,* reinforced with skeletal plates
   c. From the stone canal, water flows into a *ring canal,* which encircles the mouth

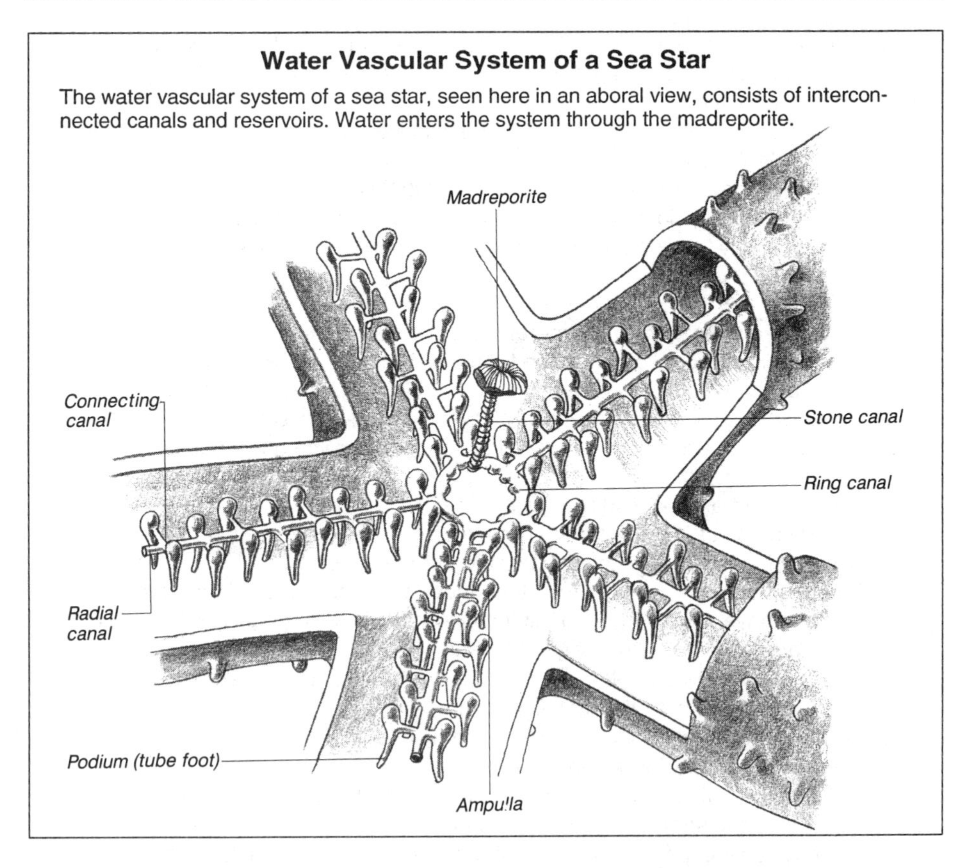

Water Vascular System of a Sea Star

The water vascular system of a sea star, seen here in an aboral view, consists of interconnected canals and reservoirs. Water enters the system through the madreporite.

d. *Radial canals* extend outward from the ring canal into the arms
e. Small *lateral (connecting) canals* join the radial canals to the tube feet

C. Locomotion

1. The podia, or tube feet, are hollow structures
    a. The outer end has a sucker
    b. The internal end has an expanded bulb-like structure, called an *ampulla*
    c. Muscle contractions in the ampulla force fluid from the main canal system into the tube foot
    d. Hydraulic pressure moves each tube foot
2. Podia can move independently, but they are coordinated to accomplish directional movement and attachment
3. Suction forces, exerted by hydraulic pressure through the tube feet, can be very powerful

D. Nutrition

1. Most echinoderms are predatory carnivores, but some are filter feeders, deposit feeders, scavengers, or browsers
2. The digestive tract is modified according to the diet

3. Sea cucumbers (class Holothuroidea) may eject their digestive tract and other organs under stress; this process is known as *evisceration,* and the lost organs are regenerated
4. Sea cucumbers also may discharge a defensive network of sticky tubules by rupturing the hindgut
   a. These structures are called *Cuverian tubules*
   b. The tubules originate from the bases of the respiratory trees
   c. The sticky tubules are ejected onto a predator, entangling it

E. Reproduction

1. Echinoderms are able to regenerate lost or damaged body parts
   a. A new individual can develop from any of the arms if a portion of the central disk remains
   b. Species that eviscerate themselves regenerate the lost organs
2. A few echinoderms (classes Asteroidea and Ophiuroidea) reproduce asexually by fission
   a. The central disk divides in half
   b. Each half of the disk forms a new individual by regeneration
3. Most echinoderms are dioecious; some are hermaphroditic
4. Except in members of class Holothuroidea, which have a single gonad, echinoderms have a pair of gonads at the base of each arm (see *Anatomy of Representative Echinoderms,* pages 131 and 132)
   a. The gametes are released through gonoducts at the base of each arm
   b. In most cases, fertilization is external
5. Development is indirect with a free-swimming larval stage
   a. Larval form varies within the echinoderm classes
   b. However, all larvae are ciliated and have bilateral symmetry (believed to be the ancestral symmetry of echinoderms)
6. Some groups, especially those in cold-water habitats, brood their eggs
   a. Fertilization is internal or occurs while the eggs are cemented to the outer body surface
   b. The eggs may be brooded within the body cavity or on the epidermis
   c. Development is direct, and the young are released as juveniles

F. Circulation, gas exchange, excretion, and osmoregulation

1. Internal transport occurs by coelomic fluid
2. The water vascular and hemal systems, which are derived from the coelom, also contribute to circulation
   a. The *hemal system,* best developed in sea cucumbers (class Holothuroidea), consists of radially arranged blood vessels and muscular pumps
   b. It functions in nutrient and gas transport
   c. Blood cells contain the respiratory pigment hemoglobin
3. Most echinoderms have no specialized respiratory organs
   a. Gas exchange occurs by diffusion across the body surface, especially through the thin walls of the papulae and tube feet
   b. Holothuroideans have specialized respiratory trees that extend through the coelom (see *Anatomy of Representative Echinoderms,* pages 131 and 132)
4. Echinoderms do not have special organs for excretion
   a. Nitrogenous wastes are removed by diffusion

    b. The primary excretory product is ammonia
5. Because they primarily are osmoconformers, echinoderms have no special mechanisms for osmoregulation; methods of osmoregulation for brackish water species are unknown

G. Nervous system and sense organs

1. The nervous system is not centralized, and cerebral ganglia are absent
2. There are three main nerve networks
    a. Each network consists of a nerve ring with radial nerves branching off into the disk and arms
    b. The networks are connected by an epidermal nerve net
3. Sensory receptors are simple
    a. Ocelli are located at the tips of the arms
    b. Sensory neurons in the epidermis are sensitive to touch, chemicals, water currents, and light

III. Classification of Phylum Echinodermata

A. General information

1. Phylum Echinodermata contains six classes
2. The classes are separated by body shape, internal anatomy, and life-style

B. Class Crinoidea

1. The two main groups of crinoids are the sea lilies and the feather stars
    a. The flower-like *sea lilies* are sessile animals that attach to the substrate with a flexible stalk on the aboral surface
    b. The *feather stars* are free-moving animals that do not have stalks
    c. Hook-like *cirri* on the aboral surface of feather stars are used for temporary attachment and assist with righting behavior; sea lily stalks also may bear cirri
2. Living stalked crinoids seldom exceed 60 centimeters in height, but fossil forms were up to 20 meters in height
3. Crinoids differ from all other echinoderms in that their oral surface is directed upward, away from the substrate
    a. The water vascular system operates on coelomic fluid; there is no madreporite
    b. Spines and pedicellariae are absent
4. Crinoids have as many as 200 arms
    a. The arms usually have many side branches, called *pinnules*
    b. Suckerless podia are distributed over the pinnules; they primarily are feeding and sensory structures
5. Crinoids are filter feeders; the arms and pinnules trap passing food particles
6. Many mobile forms are nocturnal
7. Examples of crinoids are *Antedon* and *Neometra* (feather stars) and *Ptilocrinus* (a sea lily)

C. Class Asteroidea

1. Asteroideans, commonly called sea stars, have no distinct separation between the arms and the central disk

2. They may have five arms, or multiples of five
3. The aboral surface usually is rough and spiny
   a. Pedicellariae and papulae are distributed over this surface
   b. The madreporite and a small anal opening are also located on the aboral surface (see *Anatomy of Representative Echinoderms*)
4. Most sea stars are opportunistic feeders; they act as predators or scavengers, depending on the available food resources
   a. They commonly feed on molluscs, crustaceans, polychaetes, and small fish
   b. Sea stars such as *Asterias* are economically important predators of clams and oysters
5. Many sea stars digest their prey externally; others employ internal digestion
   a. The mouth leads into a two-part stomach in the central disk
   b. The lower *cardiac stomach* is everted and secretes digestive enzymes onto the prey
      (1) This stomach is very flexible and even can be inserted between the shells of a bivalved mollusc to facilitate digestion of the prey
      (2) The liquified food then is ingested
   c. The upper *pyloric stomach* connects with digestive ceca in the arms, where digestion is completed extracellularly
6. Tube feet, controlled by the water vascular system, function in feeding and locomotion
7. Excretion by diffusion is augmented by coelomocytes
   a. The coelomocytes accumulate nitrogenous wastes
   b. The waste-carrying cells migrate to the tips of the papulae
   c. The papulae pinch off at the tips, expelling the cells and waste material
8. Examples of asteroideans are the Northern sea star *(Asterias)*, the cushion star *(Pteraster)*, the sun star *(Crossaster)*, and a blue Pacific sea star *(Linckia)*

D. Class Ophiuroidea

1. Ophiuroideans are commonly called brittle stars and basket stars
   a. Most brittle stars are small, with a central disk a few centimeters in diameter
   b. Basket stars are larger, with a central disk 4 to 5 centimeters across; the arms of basket stars are highly branched
2. In contrast to the asteroideans, ophiuroideans have a distinct separation between the arms and the central disk
   a. The arms are slender and without papulae or pedicellariae
   b. Each arm consists of a series of jointed ossicles covered with skeletal plates, or *arm shields*
   c. The articulated ossicles on the arm form a strong, flexible skeleton that has been compared to the spinal column of vertebrates
3. Locomotion occurs by movement of the arms; spines distributed along the sides of the arms increase traction
4. The tube feet have no suckers; they play a minor role in locomotion and function primarily in feeding
5. Brittle stars may be filter feeders, active predators, deposit feeders, or scavengers and many use more than one mode of nutrition; basket stars are predatory, spreading their arms in the current to catch small prey
   a. The digestive tract is reduced and confined to the central disk
   b. Digestion and absorption probably occur in the stomach

Anatomy of Representative Echinoderms

The illustrations below display internal and external structures of three representative echinoderms–sea star, sea cucumber, and sea urchin.

Sea star

Sea cucumber

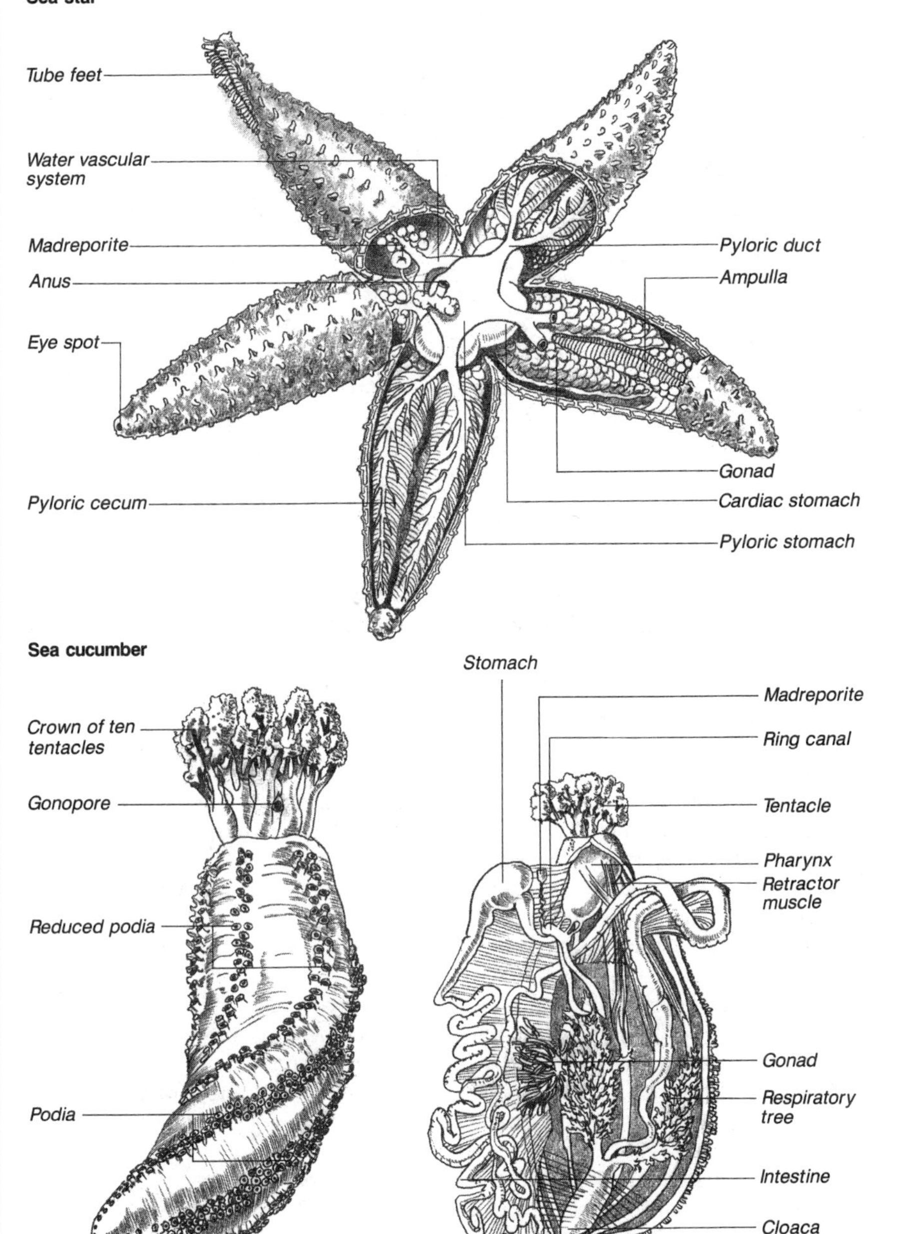

(continued)

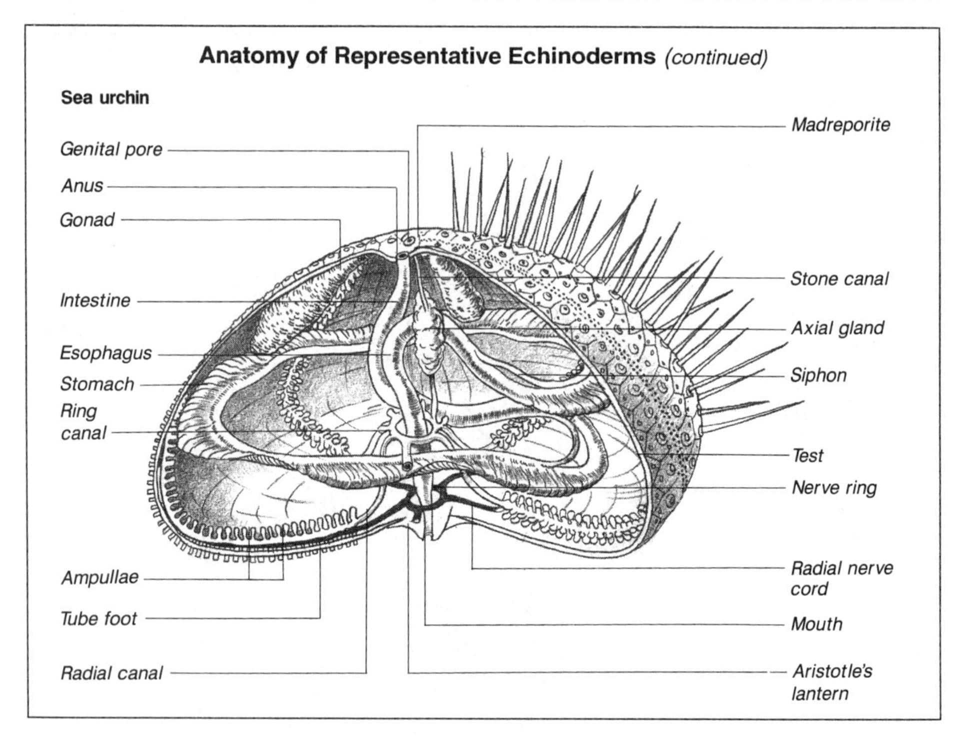

6. If captured, a brittle star can voluntarily break off its arms at any point, in a process known as *autotomy* (hence the common name, "brittle" star); the missing arm regenerates
7. Examples of ophiuroideans are the brittle stars *Ophiothrix* and *Ophiopholis* and the basket star *Gorgonocephalus*

E. Class Echinoidea

1. This class, which includes the sea urchins, heart urchins, and sand dollars, is characterized by a shell (test) that is composed of fused dermal ossicles (see *Anatomy of Representative Echinoderms*)
2. Echinoideans do not have arms, but the test reflects the pentamerous symmetry of other echinoderm groups
   a. The body surface is covered with movable spines mounted on rotating tubercles
      (1) The spines function in locomotion and defense
      (2) They also help the animal burrow into the substrate
   b. Pedicellariae are distributed over the body surface
   c. Long, suckered tube feet radiate outward from five ambulacral plates; locomotion occurs by coordinated movement of the tube feet, similar to that of the sea stars
3. Echinoideans are herbivores, detritivores, or filter feeders
4. They have a complex feeding apparatus called the *Aristotle's lantern*
   a. The lantern is composed of five calcareous teeth with associated plates and muscles

b. The jaws of the lantern are protrusible (except in sand dollars) and can be oriented at various angles
c. The teeth are used to scrape, dig, tear, or chew food into smaller pieces
5. In sand dollars and heart urchins, five areas of specialized flattened podia, known as *petaloids,* function in respiration; in sea urchins, respiration primarily is by diffusion through thin-walled podia
6. Echinoideans have four or five radially arranged gonads
a. Fertilization is external
b. Some species brood their eggs among the spines or on the petaloids (seen in sand dollars)
7. Examples of echinoideans are the purple sea urchins *(Strongylocentrotus)*, black sea urchins *(Diadema)*, heart urchins *(Meoma)*, and sand dollars *(Mellita)*

F. Class Holothuroidea

1. The approximately 1,150 species of living holothuroideans commonly are called *sea cucumbers*
2. Sea cucumbers return to bilateral symmetry as adults (see *Anatomy of Representative Echinoderms*, page 131)
a. The thick, fleshy body is elongated along the oral-aboral axis; it may be cucumber shaped (cylindrical) or worm-like
b. The mouth, with retractable tentacles, is at the anterior end; the anus is located at the posterior end
c. In most species, the skeleton is reduced to small ossicles, and the body wall is leathery
3. Most sea cucumbers are filter or deposit feeders
4. Sea cucumbers are benthic crawlers or burrowers; those that live on the surface are usually *cryptic,* seeking shelter under rocks or in crevices
5. Locomotion occurs by means of podia or through contractions of body wall musculature
6. The hemal system is well developed
7. The respiratory tree, found only in holothuroideans, functions in both respiration and excretion
8. Sea cucumbers are dioecious or hermaphroditic
a. They have a single gonad
b. Fertilization is external with indirect development
9. Examples of holothuroideans are *Thyone* and *Holothuria*

G. Class Concentricycloidea

1. Members of this class, newly discovered in 1983, are found in deep-water habitats
2. Only two species are known, *Xyloplax medusiformis* and *Xyloplax turnerae;* they commonly are called *sea daisies*
3. These small animals (less than 1 cm in diameter) superficially resemble cnidarian medusae (hence the species name *medusiformis*)
a. Arms are absent, and the podia are arranged in a ring around the edge of the body
b. The water vascular system has two ring canals, but no radial canals
c. The gut may be present or absent; there is no anus

Study Activities

1. List four basic characteristics of the echinoderms.
2. Describe the structure and operation of the water vascular system.
3. Describe how echinoderms carry out nutrition, gas exchange, and excretion.
4. Describe methods of asexual and sexual reproduction in echinoderms.
5. Create a table or chart that compares and contrasts the basic characteristics of each of the six echinoderm classes.

14

Nonvertebrate Chordates

Objectives

After studying this chapter, the reader should be able to:
- Describe the basic characteristics of chordates.
- Explain current hypotheses concerning the evolutionary ancestry of the chordates.
- Describe how the nonvertebrate chordate phyla perform their basic life functions.

I. Basic Characteristics

A. General information

1. Chordates (phylum Chordata) share many features with higher invertebrate groups, such as bilateral symmetry, eucoelomate body plan, metamerism, and cephalization
2. Chordates and echinoderms also share the deuterostome pattern of embryologic development
3. There are more than 50,000 species of chordates, both vertebrate and nonvertebrate
   a. Nonvertebrate chordates include the tunicates (subphylum Urochordata), and the lancelets (subphylum Cephalochordata)
   b. Vertebrate chordates include jawless fishes, cartilaginous fishes, bony fishes, amphibians, reptiles, birds, and mammals

B. Ecologic relationships

1. Chordates are a diverse and successful group that inhabit terrestrial, marine, and freshwater environments
2. Nonvertebrate chordates are important components of oceanic food chains
3. Both nonvertebrate (such as *Amphioxus*) and vertebrate chordates are economically important food sources for humans

C. Evolutionary relationships

1. Current evidence suggests that chordates evolved from an early echinoderm, an early hemichordate (worm-like marine animal with several chordate characteristics), or an unknown common ancestor that led to all these groups
2. Homologies among vertebrate chordates, nonvertebrate chordates, and echinoderms show a close evolutionary relationship
   a. Modern echinoderms and chordates have several structural similarities, especially in details of embryologic development (deuterostome pattern)

b. All three groups have a coelom divided into three regions and similar subepidermal organization of nervous tissues
3. Opinions differ on the nature of the ancestral form
a. Echinoderms and chordates may be "sister groups" descended from a common ancestor
b. The cephalochordates and vertebrates may have arisen by a process of neoteny (see Chapter 2, Evolution of Animal Diversity) from an *ascidian larva* (subphylum Urochordata)
c. Alternatively, one ancestral chordate may have given rise to the urochordates, while another gave rise to the cephalochordates and vertebrates
d. Most experts agree that the common ancestor, of whatever origin, was probably a benthic filter feeder with a tadpole-like larva

D. Common chordate characteristics

1. All chordates are deuterostome eucoelomates with bilateral symmetry; in some groups, the coelom is secondarily lost
2. All have pharyngeal gill pouches at some stage of development (see *Internal and External Anatomy of a Tunicate*)
a. The gill slits probably originated as an adaptation to facilitate gas exchange or filter feeding
b. Water flows into the mouth, through the pharynx, and out the gill slits
c. In primitive chordates, the gill slits function in respiration
d. In vertebrates, the gill slits may not completely perforate the pharyngeal wall; they remain as gill "pockets" that do not persist in the adult
3. Primitive chordates have a ciliated groove, called an *endostyle,* that runs along the ventral surface of the pharynx
a. The endostyle is a derived characteristic unique to chordates
b. It secretes mucus and is part of the feeding apparatus
c. Specialized endostyle cells accumulate iodine and secrete iodine-containing proteins into the mucus net; the physiologic role of these cells is not completely understood
d. The endostyle is lost in vertebrates, but the vertebrate *thyroid gland* is homologous to, and may have developed from, the iodine-binding endostyle cells
4. A dorsal *notochord* is present at some stage in the chordate life cycle
a. The notochord is a flexible, rod-like structure that provides skeletal support
b. It extends longitudinally through the body, terminating at the end of the tail, and is a point of attachment for body wall musculature
c. In some chordates, the notochord is retained in the adult form and functions in locomotion
d. In jawed vertebrates, cartilaginous or bony vertebrae encircle and then replace the notochord
5. All chordates have a single *dorsal hollow nerve cord;* invertebrates have ventral solid nerve cords
6. A *postanal tail* (projects beyond the anus) is present at some stage in the life cycle; in cephalochordates and larval urochordates, the postanal tail persists as a functional locomotor structure
7. Muscle bundles *(myotomes)* are segmentally arranged in a nonsegmented trunk
8. The digestive system is complete and usually has regional specializations
9. The circulatory system is closed, with a ventral pumping organ

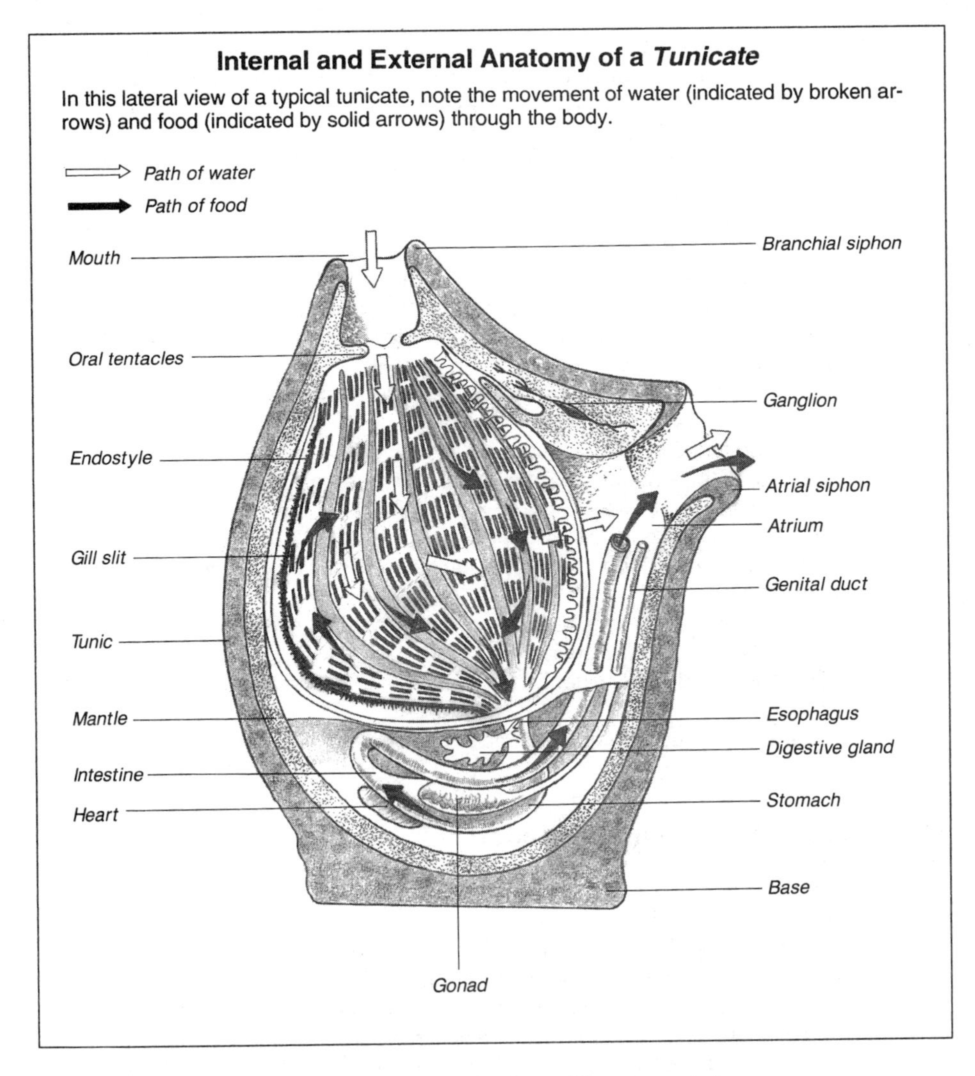

II. Basic Characteristics of Subphylum Urochordata

A. General information

1. Approximately 3,000 species of urochordates, commonly called *tunicates* and *salps,* have been described
2. Urochordates range in size from microscopic interstitial forms to individuals as long as 60 centimeters
3. Most urochordates have a spherical or tubular shape, and the body is covered by a fibrous mucopolysaccharide sheath, called a *tunic*
4. In adult urochordates, most chordate features are reduced or absent, including the notochord, tail, and dorsal tubular nerve cord

B. Ecologic relationships

1. The tunicate classes have diverse life-styles and habitats

2. All species are marine, and most are filter feeders
3. The three general groups of tunicates are categorized by body form and life-style
   a. *Solitary* tunicates are larger species that live as unattached individuals; examples include *Ciona* and *Styela*
   b. *Social* tunicates live in attached clumps, with individuals connected at the base; examples include *Clavelina*
   c. *Compound* tunicates form colonies of many small zooids attached by a common gelatinous matrix
      (1) The colonies can grow up to several meters in diameter
      (2) Examples include *Botryllus* and *Perophora*
4. The *thaliaceans* (commonly called *salps*) are similar in shape to ascidians but have a different life-style
   a. Thaliaceans are pelagic, ranging in depth from the continental shelf to 1,500 meters below the surface
   b. They are found in all seas, but are most abundant in tropical waters
   c. Individuals may be solitary or colonial and resemble social or compound ascidians in body form
   d. Examples include *Salpa* and *Thetys*
5. The *larvaceans* are unique among the tunicates in that they retain larval features in the adult
   a. Larvaceans are solitary, planktonic individuals that secrete a gelatinous sheath around their bodies
   b. The notochord, dorsal tubular nerve cord, and postanal tail are present in the adult
   c. Examples include *Oikopleura* and *Fritillaria*

III. Urochordate Form and Function

A. General information
1. Tunicates have globular or tubular bodies; sessile forms are attached at the base to the substrate (see *Internal and External Anatomy of a Tunicate,* page 137)
2. The body is covered and supported by a secreted tunic
   a. The tunic is composed of *tunicin,* a derivative of cellulose
   b. It has cellular constituents and may be considered a type of exoskeleton
3. The coelom is absent; it is replaced by an atrium or cloacal water chamber, an aqueous chamber that functions in filter feeding
4. Contractions of longitudinal and circular muscle layers pump water through the body

B. Locomotion
1. Most urochordates are sessile
2. Pelagic forms, such as salps, differ from other urochordates in body structure
   a. The buccal and atrial siphons are at opposite ends of the body
   b. This siphon arrangement directs the water stream directly through the body, moving the animal forward

C. Nutrition
1. Urochordates are filter feeders

a. Water enters through the branchial siphon and passes through the pharyngeal gill basket (see *Internal and External Anatomy of a Tunicate,* page 137)
b. The endostyle, a ciliated groove located on the ventral side of the pharynx, secretes a mucus net
c. The mucus traps organic particles, which are carried to the stomach
d. Water exits through the atrial siphon

2. Digestion is extracellular and takes place in the stomach
   a. Enzymes are secreted into the stomach by cells of the digestive tract and associated digestive glands
   b. The anus usually opens into the atrium; feces are removed via the atrial siphon

D. Reproduction

1. Most urochordates are capable of both asexual and sexual reproduction; larvaceans reproduce only sexually
2. Asexual reproduction occurs by budding; the buds can develop from a variety of body structures
3. In addition to budding, tunicate colonies engage in asexual fusion
   a. As the colonies increase in diameter, they may meet and fuse with other colonies
   b. Both interspecies and intraspecies fusion can occur
4. Most tunicates are hermaphroditic
   a. The reproductive organs are simple; in most cases, there is only a single ovary and testis
   b. Fertilization is external
5. Development usually is indirect and involves a *tadpole* larval stage
   a. The tail of the tadpole larva contains several primary chordate features
      (1) These include a dorsal tubular nerve cord, notochord, and muscle somites
      (2) Sensory receptors in the tadpole larva are better developed than those in the adult
   b. The larval stage is short, and larvae usually do not feed
      (1) Most settle to the substrate after 1 to 2 days; in some species, the larvae remain active for only a few minutes
      (2) The tail is quickly reabsorbed, and the larvae assume the adult form

E. Circulation, gas exchange, excretion, and osmoregulation

1. Urochordates have a closed circulatory system; the heart is ventral and located near the stomach
   a. Urochordates have a unique circulatory pattern, in which the blood flows first in one direction, then reverses and flows in the other; no other chordate has this alternating flow pattern
   b. The blood contains a variety of cells and transports several hormones (such as thyroxin, oxytocins, and vasoconstrictors) similar to those of vertebrates
2. Urochordates do not have specialized respiratory structures; gas exchange occurs by diffusion across the body wall
3. No specialized excretory or osmoregulatory organs are present; nitrogenous wastes are removed primarily by diffusion

F. Nervous system and sense organs

1. The urochordate nervous system is greatly reduced
2. Tunicate larvae possess a well-developed dorsal nerve cord in the tail region, which is absent in the adults

IV. Basic Characteristics of Subphylum Cephalochordata

A. General information

1. Subphylum Cephalochordata has only about 24 living species
2. Cephalochordates, commonly known as *lancelets*, are small animals, about 3 to 7 centimeters in length
3. Examples include the well-known amphioxus *(Branchiostoma)*

B. Ecologic relationships

1. Lancelets are benthic marine and brackish water organisms; they are distributed worldwide
2. Lancelets can swim, but usually burrow into the sand tail first, with the head extended for feeding

V. Cephalochordate Form and Function

A. General information

1. The cephalochordate body is elongated and laterally compressed, similar to that of a fish (see *Internal and External Anatomy of the Cephalochordate Branchiostoma*)
2. The four basic chordate characteristics are present in both larva and adult
3. The notochord functions in both structural support and locomotion
4. Caudal and dorsal fins are present
5. Segmented, V-shaped muscles, called *myotomes* or *myomeres,* run along both sides of the body
6. The coelom is reduced, and the space is occupied by body wall musculature

B. Nutrition

1. Cephalochordates are filter feeders
2. The method of feeding resembles that of tunicates
   a. Water is pulled into the mouth by the beating of cilia in the mouth and pharynx
   b. Water passes through the pharyngeal gill slits into the atrium, then exits through a ventral opening, called the *atriopore*
      (1) There may be more than 200 gill slits
      (2) The slits are reinforced by cartilaginous *gill bars*
   c. Food particles are trapped in a mucus layer secreted by the endostyle
3. Before entering the mouth, food particles are sorted in an anterior extension of the body called the *oral*, or *buccal, hood* (see *Internal and External Anatomy of the Cephalochordate Branchiostoma*)
   a. Projections called *buccal cirri,* extending from the oral hood, separate large and small particles

Internal and External Anatomy of the Cephalochordate *Branchiostoma*

In this lateral view of a typical cephalochordate, note the presence of the dorsal tubular nerve cord, notochord, pharyngeal gill slits, and postanal tail. The endostyle is seen along the ventral margin of the pharynx.

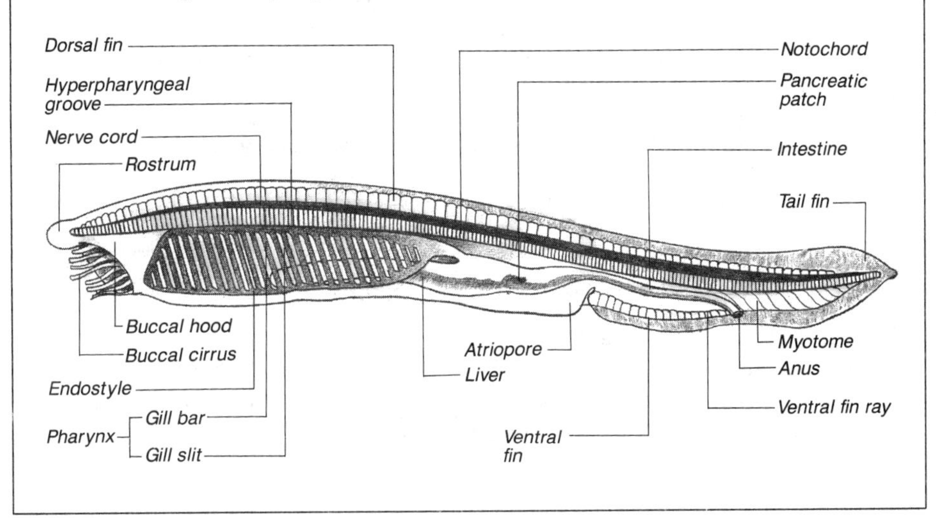

   b. Small particles continue into the mouth and are trapped in the mucus layer; large particles are rejected
4. Trapped food particles are carried to the gut for digestion
   a. Primary digestion is extracellular and occurs in the gut lumen
   b. Digestion is completed within the cells of the intestine and digestive ***cecum***
      (1) The cecum functions in lipid and glycogen storage and in protein synthesis
      (2) The vertebrate liver and pancreas may have developed from this structure

C. Reproduction

1. Cephalochordates are dioecious
   a. As many as 38 pairs of gonads are arranged along the sides of the atrium
   b. Fertilization is external
2. Development is indirect; a free-swimming planktonic larva gradually develops into an adult

D. Circulation, gas exchange, excretion, and osmoregulation

1. Cephalochordates have a closed circulatory system
   a. The pattern of blood flow resembles that of primitive fish (see Chapter 15, Subphylum Vertebrata)
   b. No heart is present; blood is propelled by contraction of muscles in the walls of the blood vessels
   c. The blood is a clear fluid that contains no cells or respiratory pigments
2. Excretion is via protonephridia, similar to the system found in primitive annelids

a. Protonephridia of cephalochordates do not appear to be homologous to those of other invertebrates
b. These analogous structures probably arose through convergent evolution

E. Nervous system and sense organs

1. The cephalochordate nervous system is simple in structure
    a. The dorsal nerve cords run the length of the body
    b. The brain is restricted to a slight enlargement at the anterior end of the nerve cord
2. Pairs of segmental nerves arise from the dorsal nerve cord
    a. The segmental nerves have dorsal and ventral roots, as do vertebrate spinal nerves
    b. The segmental nerves innervate the myotomes and other body structures

---

Study Activities

1. List six basic characteristics of chordates.
2. Discuss current hypotheses on the evolutionary ancestry of chordates.
3. Create a chart or table that compares and contrasts the basic characteristics and life-styles of the urochordates and cephalochordates.

15

Subphylum Vertebrata

Objectives

After studying this chapter, the reader should be able to:
- Describe the basic characteristics of vertebrates.
- Explain current hypotheses concerning the evolutionary ancestry of vertebrates.
- Describe how vertebrates perform their basic life functions.
- Identify and characterize the classes of subphylum Vertebrata.

I. Basic Characteristics

A. General information

1. Subphylum Vertebrata is a large and successful group; about 40,000 living species have been described
2. Vertebrates have all the basic chordate characteristics (see Chapter 14, Nonvertebrate Chordates), plus several unique derived features
3. This subphylum is named for the vertebral column (a series of compact bony elements) that surrounds the nerve cord

B. Ecologic relationships

1. All vertebrate ancestors were aquatic and probably fed on small organic particles removed from the water or the sea floor
2. Modern vertebrates display a variety of physical forms, life-styles, and specialized adaptations

C. Evolutionary relationships

1. The origin of vertebrates is still under debate, but most scientists agree on some general relationships
2. The first vertebrates probably arose from a sessile form, such as the ascidians, that had a free-swimming larva for dispersal
3. The vertebrate body organization may have arisen as an adaptation for a more active life-style
   a. This probably occurred by a process known as *neoteny,* in which the length of the larval period is gradually extended, until, eventually, sexual maturity is reached without undergoing metamorphosis
   b. The animal retains its larval form and features as a sexually reproducing adult

II. Vertebrate Form and Function

A. General information

1. Vertebrates possess all the basic chordate features at some stage in their life cycle
    a. In all vertebrates except the fish and larval amphibians, the gill slits do not perforate the body wall
        (1) They are present in the embryo as pharyngeal pouches
        (2) As development proceeds, the pouches are greatly modified or eliminated
    b. The dorsal tubular nerve cord develops into the brain and spinal cord
    c. The notochord is replaced in jawed vertebrates by a backbone of vertebrae
    d. Most vertebrates (with the exception of ape-like primates) retain the postanal tail into adulthood
2. Vertebrates are characterized by an internal skeleton of bone or cartilage (or a combination of both materials)
    a. Vertebrate skeletons are divided into two components
        (1) The *axial skeleton* includes the vertebral column and the cranium, which surrounds the brain
        (2) The *appendicular skeleton* includes the bones of the paired appendages and associated support structures
    b. Well-developed muscles are attached to the skeleton and facilitate active movement
3. The coelom is large and encloses the organs; it does not function in respiration or as a hydrostatic skeleton
4. The digestive system is complete, with complex accessory glands
5. The closed circulatory system has a ventral heart
6. Most vertebrates are dioecious; asexual reproduction does not occur

B. Adaptations for an active life-style

1. In vertebrates, paired appendages facilitate locomotion
    a. In fish, the fins facilitate balance, propulsion, and directional movement; fish can execute delicate and complex maneuvers using their fins
    b. In other vertebrates, the fins have developed into jointed limbs for movement and support on land; jointed limbs are well adapted for moving over surfaces that are not smooth or level
2. The ciliated pharyngeal gill structure is modified for respiration
    a. The internal gills are very proficient in gas exchange
    b. Respiratory efficiency far exceeds that of any invertebrate gills
    c. Oxygen is distributed around the body by a well-developed circulatory system
3. The endoskeleton of vertebrates has several advantageous features
    a. It is composed of living tissue, and thus grows with the animal, rendering molting unnecessary
    b. It permits greater body size by providing a lighter support structure
    c. It provides an increased surface area for muscle attachment (compared to exoskeletons), permitting a wider range of movements
4. Vertebrate nervous systems are highly developed and complex

III. Class Agnatha: The Jawless Fishes

A. General information

1. Ancestral jawless fishes, called *ostracoderms,* were small and heavily armored
2. Modern jawless fishes, called *agnathans,* have no body armor but share other characteristics with their fossil relatives
   a. The jaws and paired appendages are absent
   b. Gill slits are numerous, exceeding those of other living fishes
3. Agnathans have slender, elongated bodies (see *Agnanthan Body Structure*)
4. About 70 species of living agnathans, commonly called *hagfish* and *lampreys*, have been described
   a. Hagfish are entirely marine; examples are the Pacific hagfish, *Bdelostoma stouti*, and the common hagfish, *Myxine* (the word myxin means slime)
   b. Lampreys include marine and freshwater species; examples are the sea lamprey *(Petromyzon marinus)*, the Pacific lamprey *(Entosphenus tridentata)*, and the freshwater lamprey *(Entosphenus similis)*
   c. Hagfish and lampreys have many differences in body structure and life history

B. Locomotion

1. The agnathan notochord provides structural support and functions in locomotion; vertebrae are absent
2. The body is flexed from side to side to provide forward movement

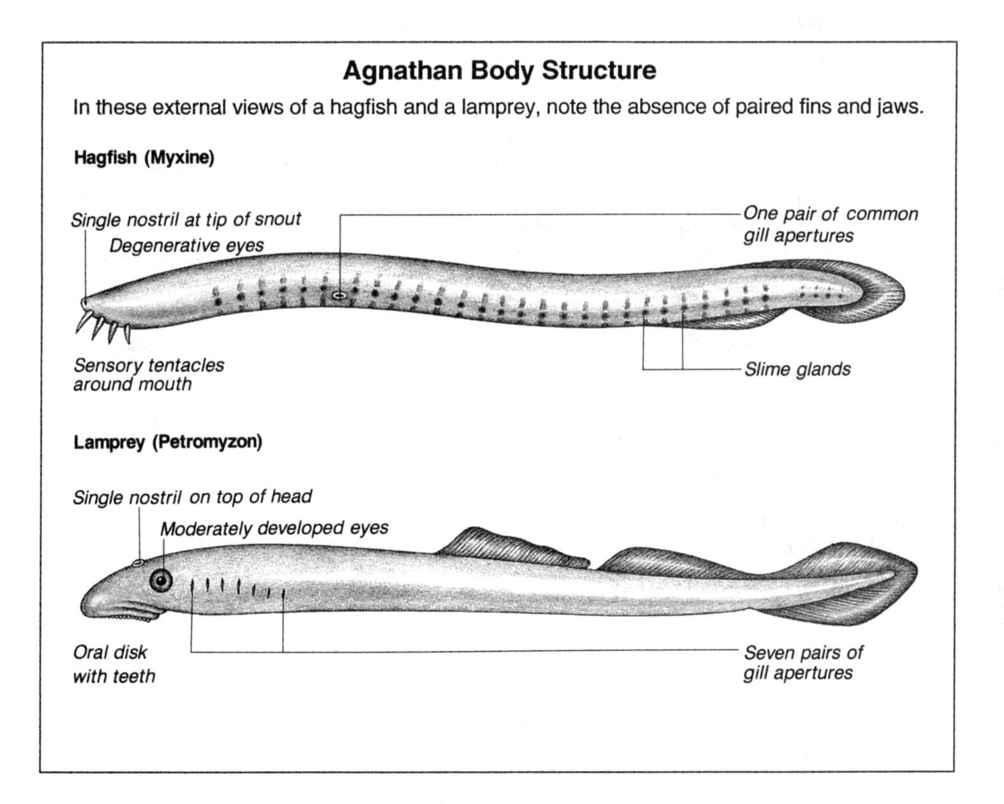

C. Nutrition

1. Larval lamprey forms are sedentary filter feeders; lampreys that feed as adults are ectoparasites, primarily of other fishes
   a. Lampreys attach to a host with the oral disk
   b. The keratin plates that cover the tongue and oral disk are used to rasp a hole in the side of the prey fish
   c. The attached lamprey feeds on host tissues and body fluids through this hole
2. Lampreys have a simplified digestive system
   a. A special adaptation of the esophagus allows food to bypass the pharynx
   b. Food travels from the mouth, through the esophagus, to the intestine; there is no stomach
3. Hagfish primarily are scavengers; their main prey are dead and dying fish
   a. Keratin plates cover the protrusible tongue
   b. The tongue is used to rasp and burrow into the body of a prey fish, usually through the gills or anus
   c. From inside the body of their prey, hagfish use their keratinized teeth to tear off bite-sized pieces of the host tissues
4. Like the lampreys, hagfish have a bypass system that allows water to pass through the pharynx and gills while the animal is feeding

D. Reproduction

1. Hagfish have both male and female sex organs, but the gonads of only one sex are functional
   a. Fertilization is external
   b. Development is direct; there is no larval stage
2. Lampreys are dioecious
   a. Development is indirect with an *ammocoete* larva
   b. The larval period is long (3 to 7 years)
3. Ectoparasitic and nonparasitic lampreys have very different life cycles
   a. All lampreys (marine and freshwater) spawn in freshwater streams
   b. The ammocoete larvae are sedentary filter feeders that burrow into the sediment with the oral hood extended to feed
   c. Adult parasitic lampreys migrate to the sea (if marine) or into large freshwater lakes
      (1) They spend their adulthood as ectoparasites of other fish
      (2) They may spend up to 2 years in the feeding phase of their adult lives
      (3) After this period, the adults return to freshwater streams to spawn and die
   d. Nonparasitic lampreys do not feed as adults; they spawn within a few months after metamorphosis and then die

E. Circulation and gas exchange

1. The hagfish heart is supplemented by up to three accessory hearts in various locations
2. Hagfish may have up to 16 pairs of gills
   a. Each gill pouch connects through one or more ducts to the pharynx
   b. The *median nostril* (a single nostril at the anterior tip of the snout) connects to the pharynx; it allows water to pass through the pharynx and gills while food is in the mouth
3. Lampreys have seven pairs of gills

a. The internal gill openings combine into a single tube connecting with the pharynx
b. With this structure, feeding does not impede the passage of water through the pharynx and over the gills

F. Excretion and osmoregulation

1. Hagfish are osmoconformers; they are the only vertebrates with body fluids isotonic to sea water
2. Hagfish also are the only vertebrates to have both pronephric and mesonephric kidneys in the adult
   a. The *pronephros* develops near the anterior end of the body
      (1) It is the first functional kidney of the embryo
      (2) In most vertebrates, this kidney degenerates early in embryonic development and is replaced by the mesonephros
   b. The *mesonephros* is the functional kidney of adult lampreys, other fish, and amphibians; it is the embryonic kidney of reptiles, birds, and mammals
   c. The *metanephros* replaces the mesonephros to become the functional kidney of adult reptiles, birds, and mammals

G. Nervous system and sense organs

1. Lampreys have well-developed sensory structures, including eyes and a lateral line system
   a. The lateral line system provides a distant touch sense
   b. Mechanoreceptors detect vibrations and water movements
   c. All fishes have some form of lateral line system
2. Hagfish have poor vision, and the lateral line system is greatly reduced

IV. Class Chondrichthyes: The Cartilaginous Fishes

A. General information

1. The cartilaginous fishes include the sharks, skates, rays, and chimaeras
   a. This ancient group represents more than 450 million years of evolution
   b. About 750 living species have been described
2. The cartilaginous fishes vary in morphology and life-style; however, all have several common features
   a. The skeleton is cartilaginous at all stages of the life cycle
      (1) The cartilaginous skeletal elements are not *ossified* (replaced by bone)
      (2) Calcium salts, which add strength to the skeleton, are sometimes deposited within the cartilage
   b. The body is covered with *placoid scales*, also called *dermal denticles*, which structurally are similar to teeth
      (1) The pulp cavity contains blood vessels and nerves
      (2) The pulp is surrounded by dentine and an outer layer of enamel
   c. Jaws are present, and the mouth is located ventrally
   d. There are 5 to 7 pairs of gill slits; an operculum (gill cover) is absent
3. Most sharks (subclass *Elasmobranchii)* are fusiform (streamlined); they range in size from less than 15 centimeters to more than 15 meters in length
   a. The caudal fin is usually *heterocercal*

(1) The tail lobes are asymmetrical, with the dorsal lobe larger than the ventral lobe
(2) The vertebral column curves upward and extends into the dorsal lobe of the tail

b. All body fins are rigid; they cannot be flexed, rotated, or collapsed

4. Sharks vary greatly in life-style, ranging from highly pelagic (with extensive geographic ranges) to resident, sedentary species
5. Skates and rays are dorsoventrally flattened
   a. The gill slits are located on the ventral side of the body
   b. The spiracles, through which water enters the gill chamber, are enlarged and located on top of the head, which prevents clogging with sediment in benthic species
   c. Some species, such as the sting ray, protect themselves with venomous spines that are usually located at the base of the tail

B. Locomotion

1. Until the cartilaginous fishes arose, most early fishes were benthic
2. Large amounts of energy are required to remain suspended above the substrate; cartilaginous fishes float by attaining *neutral buoyancy*
   a. The cartilaginous skeleton contributes to neutral buoyancy by decreasing the average density of the body
   b. Large, oil-filled livers, which are present in most active swimmers among the cartilaginous fishes, also contribute to flotation
   c. The air-filled swim bladder, which adjusts buoyancy in the bony fishes, is absent
3. Sharks and rays are heavier than water and must constantly swim forward to avoid sinking to the bottom
4. Actively swimming sharks have several adaptations for buoyancy
   a. In many species, the heterocercal tail provides lift for the posterior end of the body
   b. The large, planar pectoral fins act as airplane wings, providing lift for the anterior end of the body

C. Nutrition

1. All cartilaginous fishes are predatory carnivores, but the diet and mode of feeding vary
   a. Many species are benthic feeders on molluscs and crustaceans; they have teeth adapted for crushing
   b. Other species feed on small fish or squid, and have teeth adapted for grasping and holding
   c. The larger predators have teeth adapted for ripping and tearing chunks of flesh from their prey
2. Cartilaginous fish have a J-shaped stomach and a short intestine (see *Internal Anatomy of the Dogfish Shark Squalus acanthias*)
   a. A spiral valve within the intestine increases the surface area for digestion and absorption
   b. It also slows the passage of food through the intestine

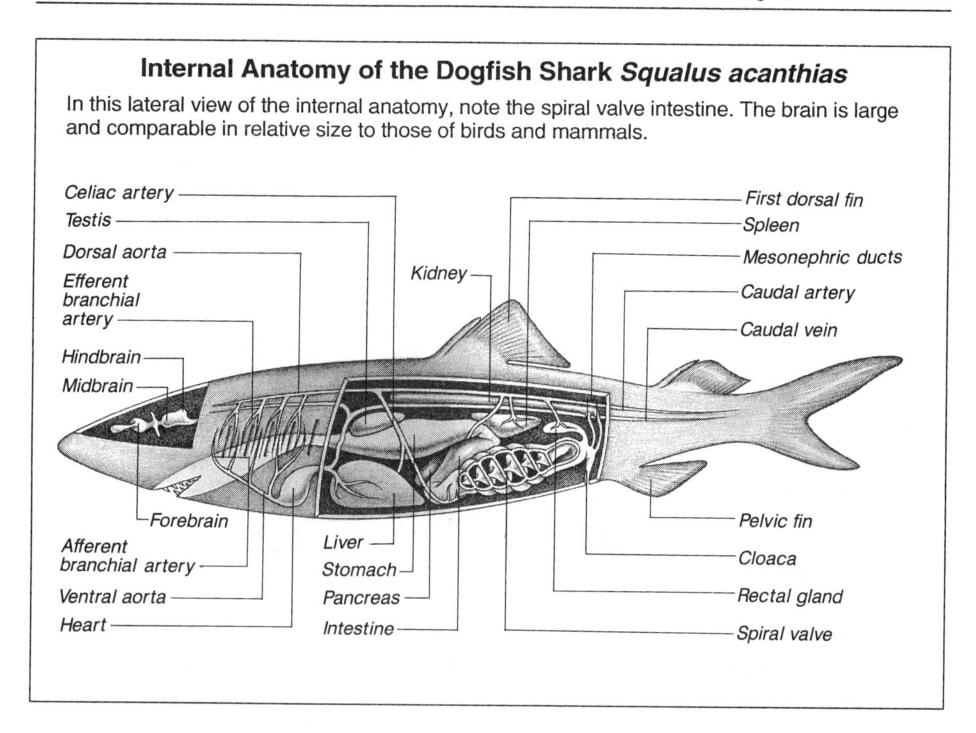

Internal Anatomy of the Dogfish Shark *Squalus acanthias*

In this lateral view of the internal anatomy, note the spiral valve intestine. The brain is large and comparable in relative size to those of birds and mammals.

D. Reproduction

1. Cartilaginous fishes are dioecious; fertilization is internal via the *clasper,* the male copulatory organ
2. Development is direct with a long gestation period (up to 2 years)
   a. Some species are *oviparous*
      (1) They lay large eggs in tough, leathery cases
      (2) The eggs are attached to the substrate to complete development
   b. Other species, including many sharks, are *ovoviviparous*
      (1) The eggs are thin shelled and hatch within the uterus
      (2) The embryos remain within the uterus to complete development
      (3) The embryos receive nutrition primarily from the yolk sac; some species secrete nutrient-rich liquids for the young
   c. Some sharks, such as members of the families Carcharhinidae (requiem sharks) and Sphyrnidae (hammerhead sharks), are *viviparous*
      (1) They produce living young, instead of eggs, from within the body
      (2) The developing embryos receive nutrition via a placenta, which is attached to an umbilical cord

E. Circulation and gas exchange

1. In cartilaginous fishes, blood is pumped from the heart, through the gills, and into arteries that supply the tissues; blood returns through veins to the heart
2. The heart has four longitudinally arranged chambers—the sinus venosus, the atrium, the ventricle, and the conus arteriosus
   a. The *sinus venosus,* the first chamber, collects blood from the venous system
   b. Contractions of the *atrium,* the large chamber dorsal to the ventricle, speed the blood flow

c. The *ventricle,* a large chamber with thick muscular walls, provides the main propulsive force for the circulatory system
d. The *conus arteriosus* has elastic walls that dampen extreme changes in pressure and provide a steady flow of blood to the gills
3. The blood has both red and white cells; the respiratory pigment is hemoglobin
4. Several species of lamniform sharks, including the mako shark *(Isurus sp.)*, porbeagle *(Lamna nasus)*, and white shark *(Carcharodon carcharias)*, maintain a body temperature as much as 12° to 18° F above that of the surrounding water
   a. A well-developed system of *retia mirabila* act as countercurrent heat exchangers
   b. The elevated body temperature may result in more efficient muscle contraction and digestion, thereby providing increased energy for swimming

F. Excretion and osmoregulation

1. The kidney is the main excretory organ in cartilaginous fish; the primary nitrogenous waste product is urea
2. Urea forms the basis of a unique osmoregulatory system
   a. Much of the urea is reclaimed from the glomerular filtrate and remains in the tissues
   b. The increased urea concentration makes the body fluids nearly isotonic to sea water
   c. The rectal gland also functions in osmoregulation, by excreting sodium and chloride ions
3. Most sharks are marine, but several species are known to migrate into freshwater rivers and lakes, where they may remain for extended periods; these species have a urea concentration about one third that of exclusively marine sharks

G. Nervous system and sense organs

1. In general, sharks possess large brains, comparable in relative size to those of birds and mammals, and are capable of complex behaviors
2. The acuity of several shark senses far exceeds that of humans, and the electromagnetic sense has no parallel in humans
   a. Sharks can see during the day and night, and most species see in color
   b. Their sense of hearing is well developed; the lateral line system is extremely sensitive to vibrations in the low-frequency range (such as the sounds of struggling fish)
   c. The sense of smell is well developed
   d. The tactile receptors on the skin also may be involved in sensing temperature and pain
3. The shark's head is covered with unique and highly specialized *bioelectrical sensory receptors* known as the *ampullae of Lorenzini*
   a. This system detects the presence and location of weak electric fields
   b. At close range, this receptor system can pick up the movements of gills and muscles, thus overcoming the camouflage and burying techniques of prey
   c. It also may function as a "compass sense" that enables sharks to navigate along the earth's magnetic field for long-distance migrations

V. Class Osteichthyes: The Bony Fishes

A. General information

1. The bony fishes are the largest and most diverse group of fishes
   a. Over 95% of all living fishes are in this class, and more than 24,000 living species have been described
   b. The bony fishes represent more than half of all vertebrate species
2. Members of this class have several common features
   a. The skeleton usually is ossified, with many vertebrae
   b. The tail usually is homocercal
      (1) The tail lobes are symmetrical, with the dorsal lobe approximately equal in size to the ventral lobe
      (2) The vertebral column ends at the base of the tail (it does not extend into the dorsal lobe)
   c. Both median and paired fins are present; they may be supported by skeletal elements, called *fin rays,* of cartilage or bone
   d. The gill slits open into a common chamber, covered by an operculum; the spiracle is absent
   e. An air-filled chamber functions in either respiration (a *lung*) or in buoyancy (a *swim bladder*)
      (1) Lungs and swim bladders are homologous organs that appeared early in bony fish evolution; the ancestral organ probably functioned as a lung similar to those of modern air-breathing fish
      (2) The swim bladder maintains neutral buoyancy at various depths; it is filled primarily with oxygen
   f. Well-developed bony scales cover the bodies of most members of this class; the number and type of scales are related to life-style
3. The three basic groups of bony fishes are the lobe-finned fish, the lungfish, and the ray-finned fish
   a. The *lobe-finned fishes* (subclass Crossopterygii) have only one living representative, the coelacanth *(Latimeria chalumnae)*; the remainder are extinct
      (1) The name "lobe fin" refers to the fleshy leg-like structure of the pectoral and pelvic fins, which enabled these fish to move about on land
      (2) Ancestral lobe-finned fishes had both lungs and gills
      (3) All terrestrial tetrapods arose from this group
   b. The *lungfishes* (subclass Dipneusti) also are remnants of an early group; only three living genera exist today
      (1) In both ancestral and modern lungfishes, air-breathing capabilities allow survival in dry conditions or in stagnant, deoxygenated water
      (2) Several modern species can survive long periods of drought; they remain dormant inside a secreted slime cocoon until the rains return
   c. The *ray-finned fishes* (subclass Actinopterygii) are the largest group of bony fishes; they have progressed through three phases in body form, all of which have living representatives
      (1) The *chondrosteans* (infraclass Chondrostei) are the most primitive ray-finned fishes; living representatives include sturgeons, bichirs, and paddlefish
      (2) Most modern bony fishes (infraclass Neopterygii) appear to have descended from one chondrostean line

(3) The *teleost* fishes (infraclass Neopterygii), the predominant modern fish, are a large and diverse group; they are found in all aquatic habitats and are specialized for every possible niche
   (a) The endoskeleton is almost completely ossified; the notochord is replaced by an ossified vertebral column (see *Internal and External Anatomy of a Bony Fish*)
   (b) The tail usually is homocercal
   (c) Scales may be either cycloid or ctenoid; both types are rounded lightweight protective structures, ctenoid scales have tiny comb-like projections (ctenii) along the posterior edge
   (d) The jaw is flexible and protrusible, with minimal attachment to skull bones
   (e) Strong, flexible fin rays form the main support structure of the fins

B. Locomotion

1. Teleost fishes are agile and can maneuver easily in the water
   a. The pelvic fins are located near the anterior end of the body, providing fine control of movements
   b. Most fins collapse to reduce drag for fast swimming

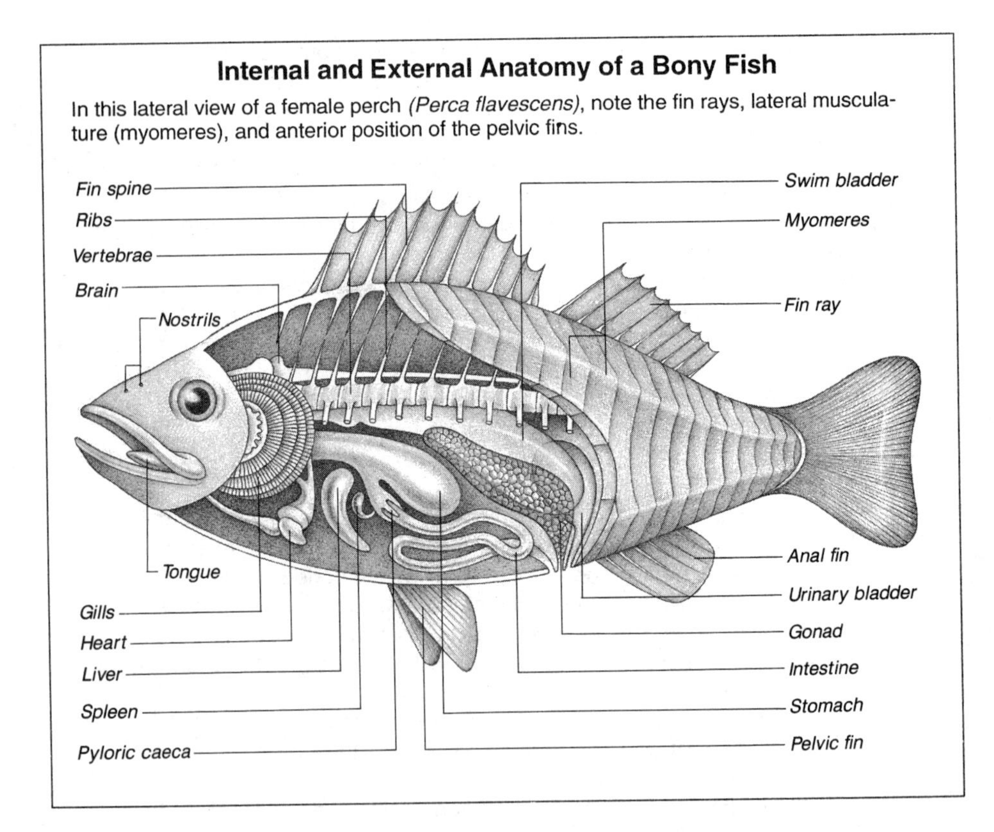

Internal and External Anatomy of a Bony Fish

In this lateral view of a female perch *(Perca flavescens)*, note the fin rays, lateral musculature (myomeres), and anterior position of the pelvic fins.

c. The fins can pivot for quick directional changes; many teleost fishes can even swim backward

2. Bony fishes move by flexing the body from side to side
    a. In *anguilliform swimming* (eel-like movement), the fish flexes the entire body; this motion is similar to that found in many sharks
    b. In *carangiform swimming,* the fish flexes less and receives a propulsive boost from the caudal fin; drag is minimized
        (1) The fastest swimming fish employ this type of swimming
        (2) Tuna and billfish, as well as lamnid sharks, are typical carangiform swimmers
3. The energy cost of movement is minimized by neutral buoyancy, which usually is achieved through the presence of a gas-filled swim bladder
    a. The volume of gas in the swim bladder regulates depth
    b. The swim bladder occupies 5% to 7% of a teleost's body volume
4. The two types of swim bladders are physostomous and physoclistous
    a. *Physostomous swim bladders* are connected by a *pneumatic duct* to the gut
        (1) This type of swim bladder is inflated by gulping air at the water's surface
        (2) It is deflated by releasing air bubbles
    b. *Physoclistous swim bladders* have no connection with the gut
        (1) They are filled and emptied by an adaptation of the circulatory system called the *rete mirabile*
        (2) The rete is an intricate countercurrent exchange system for the transfer of blood gases
            (a) Oxygen and carbon dioxide are secreted to inflate the swim bladder
            (b) Gases diffuse back into the bloodstream to empty the swim bladder

C. Nutrition

1. Teleosts have a variety of feeding modes; most are carnivores, but many are herbivores, detritivores, or omnivores
2. The shape, size, and orientation of the mouth is related to feeding habits
    a. Bottom feeders have a subterminal, downward-oriented mouth
    b. Surface feeders have a mouth that faces upward
    c. In most species, the mouth faces directly anterior; the size of the mouth and jaws is positively correlated with the size range of the preferred prey
3. Herbivores and detritivores usually have long digestive tracts with adaptations for increased surface area; carnivores tend to have shorter digestive tracts
    a. The intestine is longer than that in cartilaginous fishes and has many intestinal ceca
    b. In most bony fish species, the intestine lacks a spiral valve
4. Food is digested by enzyme and acid secretions in the stomach and intestine; nutrients are absorbed in the intestine

D. Reproduction

1. The modes of reproduction and types of breeding behavior in bony fish are diverse and complex
2. Most bony fishes are dioecious and employ external fertilization
    a. Males and females may have a similar appearance or vary widely in size and coloration

b. Many species are hermaphroditic and change sex during their life cycles; they may begin life as either males or females
c. Most hermaphrodites do not function as both male and female simultaneously
3. The basic reproductive strategy is to produce numerous (sometimes millions) of eggs and provide little or no parental care; elaborate mating rituals may precede spawning
4. Development usually is external (oviparous), but some teleosts are ovoviviparous and bear live young
5. Growth is temperature dependent
a. Temperate fish have their main growth period during the warmer months (when food is abundant and the temperature is favorable for metabolism)
b. Tropical fish grow year-round
6. The growth rate is indicated by annual rings in the scales

E. Circulation and gas exchange
1. In bony fishes, blood is pumped from the heart, through the gills, and into the arteries that supply the tissues; blood returns through veins to the heart
2. The heart has four longitudinally arranged chambers
a. The pattern of circulation and function of each heart chamber is similar to that of sharks
b. In teleost fishes, the last heart chamber is called the *bulbus arteriosus;* in contrast to the conus arteriosus of elasmobranchs and other primitive fishes, the bulbus has no valves
3. The blood has both red and white cells; the respiratory pigment is hemoglobin
4. Some pelagic fish, such as tuna and mackerel, maintain a body temperature above that of the surrounding water
a. The temperature is maintained by a system of retia mirabila
b. This system is similar to the rete system that inflates the swim bladder and similar to that found in warm-bodied sharks
5. A unidirectional flow of water passes over the gills
a. Water enters the mouth, passes the gills, and exits through the operculum
b. The water current is created by forward movement with the mouth open, or by muscular pumping
6. The gills are characterized by a large surface area with a highly efficient gas exchange surface
a. Bony or cartilaginous arches stiffen the thin *gill filaments*
b. Each filament is covered with flattened plates of epidermal tissue, called *lamellae*, which are the primary sites of gas exchange
c. The flow of water is opposite to that of blood flow, providing a countercurrent exchange system; fish can remove up to 85% of the oxygen from water passing over the gills

F. Excretion and osmoregulation
1. Bony fishes excrete most nitrogenous wastes as ammonia, which is removed primarily via the gills
2. Marine teleosts excrete excess monovalent ions (especially Na^+ and Cl^-) via specialized cells in the gills
a. These cells use an active pumping mechanism to transport ions against the concentration gradient back to the sea water

b. Excess divalent ions (such as Mg^{2+} and SO_4^{2-}) are removed by the kidneys

3. Freshwater teleosts produce a large volume of dilute urine to remove the excess water that is constantly diffusing across the gills
   a. Monovalent ions (such as Na^+ and Cl^-), also moving with the concentration gradient, are constantly being lost across the gills
   b. Gill cells use an active pumping mechanism to regain these ions, which are transported against the concentration gradient into the body
   c. This pump also maintains the ionic electrical balance (positive and negative ions) and acid/base balance and eliminates ammonia (the nitrogenous waste product)

G. Nervous system and sense organs

1. Brain volume to body weight ratios vary in bony fishes
2. Many bony fishes are capable of complex behaviors
3. The sensory capabilities of bony fishes are similar to those of cartilaginous fishes, except that ampullae of Lorenzini are absent
4. The pineal organ is sensitive to light in some bony fishes and may play a role in the timing of circadian rhythms and reproductive cycles

VI. Class Amphibia: The Amphibians

A. General information

1. All of the approximately 4,000 species of living amphibians are ectothermic
   a. They are characterized by a smooth, moist skin that contains many glands
   b. They do not have scales
2. Amphibians are primarily *tetrapod* vertebrates, meaning that they have four limbs
   a. The skeleton is largely ossified
   b. Ribs often are absent
3. Larval amphibian forms usually are aquatic and breathe with gills; adults primarily are terrestrial and breathe via lungs
4. Most amphibians have little resistance to dehydration and are found in moist habitats
5. Most terrestrial adults must return to the water to reproduce

B. Locomotion

1. Amphibian limbs are adaptations of the strong, fleshy fins of lobe-finned fishes (the probable amphibian ancestors)
2. Amphibians have musculoskeletal adaptations for terrestrial locomotion
   a. Vertebrae are heavily ossified and tightly locked together
   b. The *pelvic* (hip) and *pectoral* (shoulder) girdles are well supported by skeletal or muscular components
   c. Movement is facilitated by powerful trunk musculature

C. Nutrition

1. Adult amphibians primarily are predatory carnivores; larval forms frequently are herbivores
   a. Intestinal length reflects diet specialization
   b. The intestines of larval herbivores are longer and more highly coiled than those of adult carnivores

2. The intestine has regional specializations
    a. It usually is divided into a large and small intestine
    b. The outer opening is called the *cloaca*

D. Reproduction

1. Many terrestrial amphibians return to the water to reproduce
2. Fertilization may be internal or external
    a. Most salamanders have internal fertilization
        (1) The male drops a spermatophore (sperm packet), which is picked up by the female
        (2) Eggs are laid in the water or in moist terrestrial locations
        (3) The salamander larval stage is similar in appearance to the adult form
    b. Frogs and toads, which usually breed in water, have external fertilization
        (1) Males deposit sperm over the eggs as they are laid
        (2) The eggs are laid in the water, usually anchored to vegetation
        (3) The aquatic *tadpole* larval stage may be extended (as long as 2 to 3 years for bullfrogs), culminating in gradual metamorphosis to the adult body form
3. Amphibian metamorphosis is triggered by the hormone *thyroxine*
4. Several species of salamanders are *neotenic* (reach sexual maturity while retaining larval characters, such as gills)
    a. Some species are *obligately neotenic*
        (1) These species do not undergo metamorphosis
        (2) They spend their entire lives in the aquatic larval state
        (3) Thyroxine is produced, but metamorphosis is not initiated; receptors at the tissue level may be lacking
        (4) Examples include the mud puppy *(Necturus)*
    b. Other species are *facultatively neotenic*
        (1) Under some environmental conditions, they remain as aquatic neotenic adults
        (2) Under other circumstances, they undergo metamorphosis and become air-breathing, adult salamanders
        (3) The axolotl of the American southwest *(Ambystoma tigrinum)*) normally becomes a sexually reproducing adult as a gill-breathing larval form, but also may metamorphose into a terrestrial adult, the tiger salamander
            (a) Metamorphosis may be triggered by higher ambient temperatures or a severe drop in the pond water level
            (b) Metamorphosis may be prevented if the secretion of thyroxine is inhibited by cold temperatures (such as those present at higher altitudes)

E. Circulation and gas exchange

1. Amphibians have a *double circulation system,* which develops a high blood pressure in the brain and organs
    a. When blood flows through the lung capillaries for oxygenation, it loses pressure
    b. Oxygenated blood returns to the heart and is accelerated under higher pressure toward the tissues

c. This circulation pattern differs significantly from the unidirectional blood flow of fishes, in which blood from the gills does not return to the heart and travels under low pressure directly to body tissues

2. Amphibians have three-chambered hearts, with two atria and one ventricle
   a. The ventricle pumps blood into a forked artery that directs the blood through two circuits
      (1) The *pulmonary circuit* carries blood to the lungs and skin for oxygenation; oxygenated blood returns to the left atrium, where most of it is pumped into the systemic circuit
      (2) The *systemic circuit* distributes blood to the organs and tissues (with the exception of the lungs); deoxygenated blood returns, in veins, to the right atrium
   b. Some mixing of oxygenated and deoxygenated blood occurs in the ventricle, but a ventricular ridge, which directs blood to the two circuits, reduces mixing
3. Most amphibians have low metabolic rates and activity levels; their oxygen requirements are not as high as those of more active vertebrates
   a. Aquatic larval forms breathe via gills
   b. Adult amphibians perform gas exchange via lungs
      (1) Air is pumped into the lungs by muscular contractions in the mouth and pharynx regions
      (2) The volume of the mouth and pharyngeal cavity determines the amount of oxygen that can be pumped to the lungs; in most amphibians, the head is wide relative to the body size
4. Amphibians also use *cutaneous respiration* for gas exchange
   a. Gas diffuses across the moist skin, which is highly vascularized
   b. Some species, such as plethodontid salamanders, do not have lungs and rely on cutaneous respiration exclusively
5. In some amphibians, the lining of the mouth and pharyngeal cavity also is highly vascularized and functions in gas exchange

F. Excretion and osmoregulation

1. Amphibians have a limited ability to resist dehydration
2. Aquatic amphibians excrete metabolic wastes in the form of ammonia; they also excrete a large volume of dilute urine, which helps remove the excess water that constantly is diffusing into the body
3. Terrestrial amphibians excrete most metabolic wastes in the form of urea, which conserves water
4. A few species, such as tree frogs, excrete uric acid, the most water-conserving form of metabolic waste removal

G. Nervous system and sense organs

1. The amphibian brain is small relative to body weight
2. The sensory receptors of aquatic amphibian larval forms are similar to those of fishes; terrestrial adults, however, have adaptations for sensory reception through air
3. The eye is screened by eyelids and cleaned by tear glands; the lens is flattened
4. The ear is modified to receive sound waves through the air

VII. Class Reptilia: The Reptiles

A. General information

1. There are approximately 7,000 living species of reptiles
2. Reptiles may be aquatic or terrestrial
    a. Aquatic species are found in both marine and freshwater habitats
    b. In contrast to amphibians, reptiles often are found in dry environments
3. Reptiles are characterized by a body covered with scales
    a. The scales, which are composed of the protein keratin, reduce evaporative water loss; they are one of the adaptations that enable reptiles to live in dry habitats
    b. Reptile scales arise from the epidermis; they are not homologous to fish scales, which are bony and arise from the dermis
    c. In some species, such as alligators, the scales are retained throughout life; they continue to grow, replacing damaged areas
    d. In other species, such as snakes, the entire epidermal layer is shed and replaced periodically
4. The reptilian body varies in shape, from compact to elongated; the skeleton is highly ossified
5. Reptile jaw construction permits great biting, crushing, or grasping force; powerful muscles facilitate jaw action
6. Reptiles lay *amniotic eggs* that are protected by a dehydration-resistant, leathery shell
    a. The embryo floats and develops in the amniotic fluid within the egg
    b. Amniotic eggs allow reptiles to complete their entire life cycle on land; water is not needed for reproduction
7. Reptiles are ectothermic, but employ behavioral thermoregulation (see Chapter 3, Body Structure and Function)
    a. Body temperature is regulated by the hypothalamus of the brain (as with birds and mammals)
    b. The body temperature of some terrestrial reptiles is maintained, through behavioral thermoregulation, at a level close to that of mammals
    c. Many lizards have a *parietal eye* on the top of their heads that tracks exposure to solar radiation
8. Class Reptilia includes the turtles (order Testudines), the lizards and snakes (order Squamata), and the crocodilians (order Crocodilia)
9. The *turtles* are relatively unchanged from ancestral species
    a. They are found in terrestrial, freshwater, and marine habitats and range in size from a few inches to more than 2 meters in length and over 300 pounds in weight
    b. The body is enclosed in a two-part shell, consisting of a dorsal *carapace* and a ventral *plastron*
        (1) The shell is fused with the ribs and vertebrae of the skeleton and cannot be removed
        (2) Some species have a long, flexible neck that can be withdrawn into the protection of the shell
    c. Order Testudines includes such diverse species as the green and leatherback sea turtles, the giant Galapagos tortoise, and the small red-eared turtle (a common household pet)

10. The *lizards* and *snakes* are the largest and most successful reptile group; about 95% of all living reptiles belong to this order
    a. Lizards occupy many diverse terrestrial and aquatic habitats
        (1) Body form varies and reflects the demands of the habitat
            (a) Adaptations include prehensile tails, expanded adhesive toe pads, protrusible sticky tongues, and cryptic coloration
            (b) In some species, limbs are absent
        (2) A few lizards, such as the Gila monster of the southwestern United States, inject neurotoxic venom (which primarily acts on the nervous system) when they bite; the venom may cause respiratory failure or blindness
        (3) The eyelids are movable, and most species have external ears
        (4) Examples include geckos, chameleons, and iguanas
    b. Snakes have elongated, limbless bodies
        (1) Both the pectoral and pelvic girdles are absent (or persist only as skeletal remnants)
        (2) The many compact, broad vertebrae permit rapid, side-to-side movements
        (3) Snake eyes are covered with a transparent cap, called a *spectacle;* there are no movable eyelids
        (4) Most species do not have external ears
        (5) Poisonous snakes inject venom through grooved or hollow fangs
            (a) Vipers, such as rattlesnakes, have long hollow fangs that articulate with the upper jaw; the fangs are folded against the roof of the mouth when the mouth is closed and extend automatically when the mouth is opened
            (b) Cobras, coral snakes, and kraits have short hollow fangs that always are erect
            (c) Other snakes, such as the African boomslang and the vine snake, have grooved fangs at the back of the mouth
        (6) Most snake venoms have both neurotoxic and hemolytic factors; hemolytic venom acts on the red blood cells and blood vessels, causing widespread hemorrhages
11. The *crocodilians* (crocodiles and alligators) include the largest living reptiles; crocodiles can attain more than 5 meters in length and 500 pounds in weight
    a. Most species are found in freshwater habitats; a few species can tolerate salt water
    b. Crocodilians have dorsoventrally flattened bodies with long, laterally compressed tails
        (1) Ears, eyes, and nostrils are located on the top of the head, and protrude when the animal is submerged
        (2) The ears and nose can be closed with valves that prevent water from entering
    c. All crocodilians are predatory carnivores; the jaws are strong and armed with numerous sharp teeth
    d. Unlike most reptiles, crocodilians make audible vocalizations; vocal sacs on the sides of the throat are filled with air to produce loud bellows

B. Locomotion

1. Reptiles have stronger skeletal support and more efficiently designed limbs than amphibians
2. Limbed reptiles usually have two pairs of limbs with five toes each
3. Limbless reptiles have unique locomotor patterns
   a. In *undulatory locomotion*, S-shaped loops of the trunk form behind the head and push against the substrate to move the animal forward
   b. In *rectilinear locomotion*, ventral scales push against the substrate to slowly move the animal forward (in a straight line)
   c. *Concertina locomotion* often is used by climbing snakes
      (1) The body is folded like an accordion and the posterior end anchors around a branch or in a crevice
      (2) The anterior end releases and reaches forward
      (3) When a new anchor point is attained, the rear end releases and is pulled forward

C. Nutrition

1. Most reptiles are predatory carnivores, but some species, such as tortoises and some lizards, are herbivores
2. Because reptiles are ectothermic, metabolic heat is not used to maintain body temperature
   a. A reptile's food requirements, therefore, are lower than those of a comparably sized bird or mammal
   b. Low energy needs allow reptiles to live in regions with scarce food resources, such as deserts

D. Reproduction

1. Reptiles are dioecious, and fertilization occurs internally
   a. Sperm from paired testes pass, via the vas deferens, into the male copulatory organ (penis)
   b. The penis is formed by evaginations of the cloacal wall
2. Most reptiles are oviparous
   a. They lay shelled eggs in constructed nests, in excavated holes, or under rocks and logs
   b. Crocodilians exhibit extended parental care of the young
3. Some snakes are ovoviviparous and a few are truly viviparous, nurturing the young with a placenta-like organ
   a. Snakes can store sperm after copulation
   b. Females may lay many clutches of eggs after one mating

E. Circulation and gas exchange

1. As with amphibians, reptiles have a three-chambered heart with a double circulation pattern; the reptile system, however, is more efficient and maintains a higher blood pressure
   a. Reptiles have a *ventricular septum* (or wall) that partially separates the two sides of the ventricle
   b. This minimizes the mixing of blood in the single ventricle, providing two functionally separate blood circuits
   c. Crocodiles have a complete septum in the ventricle, providing a four-chambered heart similar to that of birds and mammals

2. Reptiles breathe almost exclusively via lungs
   a. The reptile lung is a *negative pressure system*
   b. Air is pulled into the lungs by changing the volume of the thoracic cavity (in contrast, amphibians push air into the lungs by muscular contractions in the mouth and pharynx regions)

F. Excretion and osmoregulation

1. Excretion and osmoregulation take place via the kidney
2. Nitrogenous wastes are excreted primarily as uric acid
   a. This excretory product, combined with reabsorption of excretory water by the kidney tubules, allows adult reptiles to conserve water
   b. It also facilitates life in desert regions

G. Nervous system and sense organs

1. The reptile nervous system is more advanced than that of amphibians
2. The brain is small relative to body size, but the cerebrum is more highly developed
   a. The enlarged cerebrum is associated with expansion in the breadth and complexity of the behavioral repertoire
   b. Among the vertebrates, crocodilians have the first true cerebral cortex
3. Snakes have special sense organs that are used to locate prey
   a. The tongue is a tactile and olfactory organ
   b. The roof of the mouth contains sensitive olfactory receptors, called *Jacobson's organs*
      (1) The tongue picks up odor particles as it moves in and out of the mouth
      (2) The tips of the forked tongue are inserted into the Jacobson's organs and placed against the sensitive epithelial lining for odor reception
   c. Pit vipers have heat-sensitive organs, called *pit organs,* between the nostrils and the eyes
      (1) The pit organs are acutely sensitive to temperature differences
      (2) This sense is used to track and strike warm-blooded prey

VIII. Class Aves: The Birds

A. General information

1. Birds are ectothermic tetrapod vertebrates
   a. More than 8,600 species have been identified
   b. The largest group (60% of all bird species) is the perching songbirds
2. Evidence of the reptilian ancestors of birds are the scales that cover the legs and feet and the amniotic eggs that birds lay
3. Birds are characterized by the presence of *feathers,* which cover most of the body
   a. Feathers are light, but very strong
      (1) These nonliving structures are composed of keratin, which is produced by the epidermis (homologous to reptile scales)
      (2) Because bird feathers are not composed of living cells, and consequently cannot be repaired, they are replaced in periodic *molts*
   b. Adult birds are covered primarily with *contour feathers*
      (1) The feather is supported by a hollow shaft, terminating in a *quill* that emerges from the skin follicle

(2) The shaft bears a *vane* composed of numerous side branches, called *barbs*
(3) Each barb has tiny hooked branches, called *barbules,* along the side; the barbules of adjacent barbs hook together to form a coherent whole
(4) If the barbs separate, the bird can *preen* the feather with its bill and re-align the hooks
(5) Enlarged contour feathers on the arm, hand, and tail are called *flight feathers*

c. Young birds are covered with soft feathers called *down*
(1) Down feathers have a reduced shaft with long, fluffy barbs; they provide excellent insulation
(2) Down is found beneath the contour feathers of aquatic birds, such as ducks and geese

4. Like the feathers, the bird skeleton is light, but strong; the pectoral limbs are modified into *wings*
5. The jaws are modified into a horny *beak;* teeth are absent

B. Locomotion

1. Birds have various modes of locomotion, including flying, swimming, and walking
2. Several structural adaptations increase power while minimizing weight
   a. Birds have extremely light, strong skeletons
      (1) The bones are thin walled and filled with air cavities
      (2) Supporting *struts* traverse the hollow spaces, providing strength with little added weight
      (3) The skeleton of a frigate bird, which has a 7-foot wingspan, weighs only 4 ounces
   b. The vertebral column and appendicular skeleton is specialized for flight
      (1) Many vertebrae are fused, providing a rigid support structure
      (2) A long, thin keel, which runs along the midventral surface of a greatly enlarged sternum, is the attachment site for the powerful flight muscles
   c. Hand and finger bones are greatly modified and reduced in number; some are fused to provide the support framework for the wing
   d. Bird wings function as airfoils to provide lift for flight
3. Not all birds can fly
   a. There are four orders of flightless birds, including the ostriches, rheas, emus, and kiwis
   b. Other orders may have flightless species as well

C. Nutrition

1. Many bird species are carnivores; others eat fruits, nuts, and seeds
2. Birds have a high metabolic rate and consequently select high-calorie diets
3. Their beaks are adapted for specialized feeding modes, such as water straining (ducks), nut cracking (parrots), meat ripping and tearing (eagles and vultures), and fish spearing (herons); teeth are absent
4. The digestive system has several chambers in which food is crushed and ground before moving to the intestine
   a. The lower end of the esophagus may be enlarged into a crop, where food is stored and softened by water
   b. The stomach secretes digestive enzymes

c. The gizzard (an extension of the stomach) is lined with horny plates that grind the food
(1) Many species swallow small stones or rough objects that remain in the gizzard and assist with the grinding process
(2) In many carnivorous species (such as owls), the gizzard is reduced
(a) In these species, the gizzard traps the indigestible parts of prey organisms and prevents this material from entering the intestine
(b) This material (such as fur or bones) is regurgitated as pellets
5. Digestion and absorption of nutrients is completed in the intestine

D. Reproduction

1. The testes of male birds are reduced and nonfunctional for most of the year; they become greatly enlarged during the breeding season
a. There is no penis or other intromittent organ
b. During copulation, the male aligns his cloaca with that of the female for sperm transfer
2. In female birds of most species, only the left ovary and oviduct develop (an adaptation for weight reduction)
a. As with testes, the ovary becomes enlarged and functional only during the breeding season
b. Fertilization occurs in the oviduct before the shell and other components (such as albumin) are added to the egg
3. All birds are oviparous, laying amniotic eggs with calcareous shells
4. Mating behavior and patterns of parental care are diverse and complex

E. Circulation and gas exchange

1. Birds have a double circulation system, similar to that of amphibians and reptiles, but with no mixing of oxygenated and deoxygenated blood
2. Birds have a four-chambered heart, with two atria and two completely separated ventricles; blood received by the tissues has a high oxygen content, because the oxygenated and deoxygenated blood do not mix in the heart
3. Because birds are endothermic, they require more oxygen than a comparably sized ectotherm; increased metabolism is used to generate body heat and maintain body temperature
4. Birds have a unique respiratory system, adapted to meet the high oxygen requirements of flight
a. Birds have a *unidirectional* flow of air through the lungs; other vertebrates have a *bidirectional* air flow pattern (air is inhaled and exhaled through the same passageway)
b. The bird lung is modified into a system of tiny channels called *parabronchi;* air flows through the parabronchi in only one direction
c. The main bronchi connect the lungs to a system of nine air sacs distributed throughout the body
5. Two cycles of inspiration and exhalation are required to move one volume of air through the respiratory system
a. In the first inspiration, air is drawn through the main bronchi to the posterior air sacs
b. During the first exhalation, the posterior air sacs are compressed, forcing air into the parabronchi, where oxygen diffuses into the capillaries

c. During the second inspiration, air from the parabronchi travels to the anterior air sacs
d. During the second exhalation, air in the anterior air sacs is exhaled to the outside

6. During inhalation, both the anterior and posterior air sacs expand, but they receive different types of air
    a. The posterior air sacs receive oxygen-rich air from the atmosphere
    b. The anterior air sacs receive deoxygenated air from the lungs
7. During exhalation, anterior and posterior air sacs expel air to different locations
    a. The posterior air sacs send oxygenated air to the parabronchi
    b. The anterior air sacs expel deoxygenated air to the outside
8. The lungs receive fresh air during both inhalation and exhalation, providing a constant stream of oxygenated air for distribution to the tissues
9. Since air flows through the lungs in only one direction, the air at gas exchange surfaces has a higher oxygen content than that found in the blind air sacs (alveoli) of other vertebrate's lungs
    a. Birds can obtain adequate oxygen even when flying at high altitudes
    b. At very low partial pressures of oxygen (those that can barely support movement in mammals), birds can fly and vocalize

F. Excretion and osmoregulation

1. Birds have large, paired metanephric kidneys with many thousands of kidney tubules; they do not have a urinary bladder, possibly as a weight-saving adaptation
2. Birds have two excretory mechanisms for water conservation
    a. Some kidney tubules have loops of Henle, which reabsorb water
    b. Most nitrogenous wastes are excreted as uric acid
        (1) The minimal water used to transport uric acid and other excretory wastes (less than 3 milliliters of water per gram of uric acid) is reabsorbed in the cloaca
        (2) Uric acid is excreted as a white, crystalline solid mixed with the feces
            (a) It is more than 3,000 times as concentrated as plasma
            (b) Even the best water-conserving desert mammals excrete urea, which does not exceed 25 times the plasma concentration
3. Marine birds have salt glands (similar to those found in marine reptiles) located above each eye; they excrete a highly concentrated sodium chloride solution through nasal openings

G. Nervous system and sense organs

1. Birds have well-developed brains and nervous systems
2. Most birds have limited senses of smell and taste
3. Hearing and vision are excellent, surpassing those of all other animals

IX. Class Mammalia: The Mammals

A. General information

1. Mammals are endothermic tetrapod vertebrates
2. Like the birds, mammals evolved from reptilian ancestors; mammals, birds, and reptiles share many structural similarities

3. Approximately 4,500 species of mammals have been described
    a. They are diverse in size, shape, and life-style
    b. They range in size from shrews, only a few grams in weight, to the great whales, which can exceed 90,000 kilograms
4. Mammals are characterized by the presence of *hair*
    a. Hair, like feathers, is composed of keratin
    b. Hair and subcutaneous fat form insulating layers for the maintenance of body temperature
5. The skin of mammals contains four basic types of secretory glands
    a. *Sweat glands* function in temperature regulation; they secrete a watery fluid that evaporates and cools the body
    b. *Scent glands* are found at various locations on the body; they are used for intraspecies communication (such as territorial marking, warning, and reproduction)
    c. *Sebaceous glands* are associated with hair follicles; they secrete an oily lubricant for skin and hair
    d. *Mammary glands* produce milk to feed the young
6. The teeth are differentiated and adapted for chewing different types of food
7. A muscular diaphragm separates the thoracic and abdominal cavities
8. The brain is large in proportion to body size; mammals are capable of complex learned behaviors
9. In most species, the young develop within a placenta

B. Locomotion

1. Mammals employ various methods of locomotion, including walking, swimming, and flying
2. Mammals have limbs and feet that are adapted for various life-styles
    a. In bats, for example, the wings are supported by four greatly elongated fingers and covered by a leathery membrane
    b. In marine mammals, the limbs are modified into flippers and fins

C. Nutrition

1. Mammals exploit diverse food resources
    a. They have many morphologic and physiologic specializations related to feeding
    b. Many mammals have specialized teeth and mouthparts suited to their diet
2. Digestion occurs primarily in the stomach
    a. The small intestine is the principal site of nutrient absorption; the inner lining is covered with numerous small folds, called *villi,* which increase the surface area for digestion and absorption
    b. Water and ions are reclaimed in the large intestine
3. In general, herbivorous mammals have long digestive tracts because plant cellulose is difficult to digest
    a. Symbiotic bacteria in the gut help to digest the cellulose
    b. In some species (horses, rodents, elephants, and others), vegetation is fermented and digested in intestinal ceca
    c. Other species (primarily rabbits and small rodents) eat their fecal pellets, giving the intestinal bacteria a second opportunity for digestion
    d. Ruminants (such as cattle and sheep) have a four-chambered stomach that functions in bacterial digestion and absorption

D. Reproduction

1. Most mammals have defined mating seasons
   a. In many species, females undergo a periodic reproductive cycle, called an *estrous cycle;* females mate only at a specified time of this cycle, referred to as *estrus* or *heat*
   b. In other species, such as humans, females can mate and conceive at any time during the year
2. The three main groups of mammals are classified by mode of reproduction
   a. The *monotremes* are oviparous
      (1) As with reptiles, monotremes have only a single opening, the cloaca, for excretion and reproduction
      (2) There are only two living representatives of this group—the duck-billed platypus and the echidnas (spiny anteaters); both are endemic to Australia
   b. The *marsupials* have a brief period of internal development (from 13 to 35 days)
      (1) Newborn marsupials are small and poorly developed
      (2) Development is completed within the mother's marsupium, or pouch; this stage of development can last for several months
      (3) Embryos attach to nipples in the pouch and feed on the mother's milk
      (4) Marsupials can become pregnant soon after giving birth, but the embryo enters a dormant stage called *diapause;* when the first offspring leaves the pouch, development in the second embryo resumes
   c. In *placental* mammals, the embryo completes development within the uterus
      (1) The placenta, a structure composed of both fetal and maternal cells, surrounds the embryo
      (2) The length of the gestation (developmental) period varies with the species; in general (with some important exceptions), the larger the mammal, the longer the gestation period
      (3) The condition of the young at birth is variable, ranging from helpless to agile and active
      (4) Litter size and frequency typically is inversely correlated with body size; small rodents, for example, produce several litters per year, with many young in each litter

E. Circulation and gas exchange

1. Mammals have a double circulation system similar to that found in birds
   a. They have a four-chambered heart, with two atria and two ventricles
   b. The blood contains both red and white cells; the red blood cells are not nucleated
2. As with birds, mammals are endothermic; their high metabolic rate requires a sizable amount of oxygen
3. Respiration occurs via lungs; the alveoli (blind air sacs) are the primary sites of gas exchange
4. A horizontal panel called the *palate* separates the air and food passages in the mouth and pharynx; a mammal can breathe and eat at the same time

F. Excretion and osmoregulation

1. Excretion and osmoregulation occur via kidneys

a. The kidney can reabsorb a high percentage of the water that passes through it; consequently, the urine is very concentrated
b. This high absorptive capacity is one adaptation that allows mammals to live successfully in desert regions
2. Most nitrogenous waste is excreted as urea

G. Nervous system and sense organs
1. Mammals have a well-developed brain and sensory receptors
a. The enlarged cerebrum contains many sensory and motor centers
b. Mammals are capable of complex, learned behaviors
2. Most species have acute senses of hearing, sight, and smell
a. Bats, for example, have special senses for navigation and prey location; this system of navigation is called *echolocation* or *sonar*
(1) The bat emits short ultrasonic pulses (above 30,000 Hz) that bounce off objects in the environment
(2) The bounced pulses are received as echoes by the bat's large ears
(3) Through these echoes, the bat can locate and capture prey while avoiding objects in its flight path
b. Whales and dolphins use a similar echolocation system to negotiate their marine environment

Study Activities

1. List four basic characteristics of vertebrates.
2. Compare and contrast vertebrate and nonvertebrate chordates.
3. Identify and characterize the seven classes of vertebrates.
4. Discuss how the various vertebrate classes carry out nutrition, circulation, gas exchange, and excretion.
5. Describe methods of reproduction in vertebrates.

16

Vertebrate Physiology

Objectives

After studying this chapter, the reader should be able to:
• Understand the structure and function of the circulatory system.
• Understand mechanisms of temperature regulation, excretion, and osmoregulation.
• Understand the structure and function of the nervous system.
• Understand the structure and function of the endocrine system.
• Understand the structure and function of the immune system.

I. The Circulatory System

A. General information

1. All vertebrates have high-pressure, closed circulatory systems
2. Intracellular and extracellular fluids differ in the type and amount of dissolved solutes
    a. All fluids are primarily water
    b. From 70% to 90% of an animal's body weight is water

B. Composition of blood

1. Blood is composed of plasma and formed elements (red and white blood cells and platelets)
    a. Red blood cells transport oxygen to cells
    b. White blood cells function in immunity and defense
    c. Platelets function in coagulation (blood clotting)
2. The plasma comprises about 55% of the total blood volume in humans
    a. About 90% of the plasma is water
    b. The remaining 10% is dissolved solutes, including plasma proteins, amino acids, glucose, respiratory gases, inorganic ions, antibodies, hormones, metabolic wastes, and other substances
3. The main plasma proteins are albumins, globulins, and fibrinogens
    a. *Albumins* help to maintain the osmotic pressure of the blood
        (1) They keep the plasma in osmotic equilibrium with the body cells
        (2) They also function as carrier proteins; for example, bilirubin, a product of the metabolic breakdown of hemoglobin, is carried by albumin
    b. *Globulins* (such as gamma globulins) are part of the immune system response; they also function as carrier proteins
    c. *Fibrinogen,* a large protein, functions in blood clotting

C. The mammalian heart

1. The mammalian heart has four chambers; circulation of oxygenated and deoxygenated blood is completely separate
2. One-way valves in the heart ensure a unidirectional blood flow
3. The heart rate depends on the animal's body mass and metabolic level
   a. The heart rate of ectotherms is lower than that of endotherms of equal weight; for example, a codfish heart beats about 30 times a minute, while a similarly sized rabbit heart beats about 200 times a minute
   b. Small animals have higher heart rates than larger animals; the smallest mammal, a shrew, has an average heart rate of 800 beats/minute; a mouse, 400 beats/minute; a cat, 125 beats/minute; a human: 70 beats/minute; and an elephant, 25 beats/minute

II. Homeostasis

A. General information

1. Vertebrates have complex systems to maintain homeostasis—a stable internal environment regardless of varying external conditions
2. Homeostatic functions regulate body temperature, chemistry, removal of metabolic wastes, and osmotic concentrations of body fluids

B. Temperature regulation

1. Endothermic mammals maintain their body temperatures between 36° and 38° C; bird body temperatures range from 40° to 42° C
2. Temperature is maintained by balancing heat production with heat loss
   a. Heat is produced by metabolic reactions and muscular activity; a large percentage of the daily calories consumed by endotherms is used to generate metabolic heat
   b. Heat is lost by conduction from exposed body surfaces and evaporation of water
3. Endotherms have morphologic and physiologic adaptations to control both heat production and heat loss; similar adaptations prevent excessive heat buildup in hot weather
   a. Fur thickens in winter; in cold climates, mammals grow a thick layer of insulating underhair
      (1) Special muscles lift hairs or feathers, trapping an insulating layer of air next to the skin
      (2) Aquatic mammals have a thick insulating layer of fat called *blubber*
   b. Blood vessels in the skin and extremities constrict, minimizing conductive heat loss; in arctic mammals, a countercurrent exchange system returns heat to the body core
   c. Shivering and increased metabolic rate also help to counteract the effect of low environmental temperatures
   d. Small birds and mammals (such as hummingbirds and bats) conserve energy through *daily torpor,* a process by which they allow body temperatures to drop for short periods when they are inactive
   e. Other small mammals (such as rodents) may *hibernate;* during these lengthy dormant periods, the animal's body temperature and metabolic rate are minimal

f. Larger mammals (such as bears) sleep in winter, but their body temperature does not decrease; this is not true hibernation

C. Excretion and osmoregulation

1. The kidney regulates the composition of body fluids and removes nitrogenous wastes
   a. When fluid intake is high, the kidney excretes a dilute urine (reabsorbing salts and eliminating excess water)
   b. When fluid intake is low, the kidney reabsorbs water and forms a highly concentrated urine
2. The *nephron* is the functional unit of the kidney; each kidney contains up to 1 million nephrons
   a. The nephron receives blood through a small, ball-like network of capillaries called the *glomerulus*
   b. *Bowman's capsule,* a cup at the end of the renal tubule, surrounds the glomerulus
   c. The glomerulus and Bowman's capsule together are called the *renal corpuscle*
      (1) The glomerulus and Bowman's capsule have very thin walls
      (2) Small dissolved molecules are filtered from the blood and forced into the Bowman's capsule by glomerular blood pressure
   d. From the renal corpuscle, fluid passes into the *proximal convoluted tubule;* proximal tubule cells actively reabsorb almost all the glucose, vitamins, amino acids, and many electrolytes present in the fluid
   e. The proximal convoluted tubule leads to the *loop of Henle*
      (1) The loop of Henle has two parallel sections, the descending and ascending limbs; fluid in the descending limb flows in the direction opposite to that in the ascending limb
      (2) The walls of the ascending limb are impermeable to water
         (a) Sodium and chloride ions are pumped out of the ascending limb into the interstitial fluids
         (b) These ions and urea (which leaks out of the lower part of the collecting tubules) increase the osmotic concentration of the fluids surrounding the descending limb, drawing water out
   f. Because salt is removed in the loop of Henle, urine entering the *distal convoluted tubule* is dilute (isotonic or slightly hypotonic to plasma)
      (1) Additional salts are reabsorbed in this tubule
      (2) Other wastes, including ions and organic molecules, are secreted into the urine by the distal tubule cells
   g. The urine then passes through the *collecting duct*
      (1) The dilute urine is hypotonic to the fluids surrounding the collecting duct
      (2) Water diffuses out of the collecting duct, producing a concentrated urine that is hypertonic to plasma
      (3) The final urine concentration depends on the permeability of the walls of the collecting duct
         (a) Antidiuretic hormone (ADH or vasopressin), released by the pituitary gland, increases the permeability of the collecting duct
         (b) When the body is dehydrated, more ADH is produced, thereby increasing water reabsorption

h. The concentrated urine passes from the collecting duct into the *renal pelvis* for excretion

III. The Nervous System

A. General information

1. The vertebrate nervous system consists of the brain and spinal cord *(central nervous system)* and the nerve processes connecting the central nervous system to other parts of the body *(peripheral nervous system)*
2. The *neuron* is the basic unit of the nervous system
   a. The neuron consists of a cell body with cytoplasmic extensions called *processes;* groups of processes form nerves
   b. *Axons* are processes that transmit impulses away from the cell body
   c. *Dendrites* conduct impulses toward the cell body

B. Transmission of a nerve impulse

1. Nerve impulses within a neuron are transmitted electrically
2. Nerve impulses are propagated on an all or none basis
   a. The stimulus must exceed a certain threshold level of strength to activate the nerve and trigger transmission of an impulse
      (1) Nerve impulses are the same in all neurons
      (2) The rate of transmission does not vary with the magnitude of the stimulus
   b. Strong stimuli may initiate several transmissions in succession
3. The *resting potential* describes a nerve cell at rest
   a. It is caused by the unequal distribution of ions on both sides of the cell membrane
   b. When not transmitting an impulse, the nerve cell membrane is selectively permeable to potassium, which diffuses through ion-specific channels in the membrane; the sodium and chloride channels in the membrane are closed
      (1) Potassium ions diffuse out of the cell, following the concentration gradient
      (2) The negatively charged ions and molecules inside the cell cannot follow
      (3) The escaping potassium ions give a positive charge to the outside of the cell membrane
      (4) The increasing positive charge on the outside of the cell membrane stops the outward diffusion of potassium while many potassium ions still remain inside the cell (resting potential is achieved)
   c. The resting potential also is maintained and balanced by sodium-potassium pumps
      (1) The pump mechanism is an active transport system requiring energy from ATP
      (2) It consists of transport proteins in the cell membrane
      (3) The pump action transports potassium ions into and sodium ions out of the cell
4. The *action potential* describes the electrical transmission of a nerve impulse
   a. It is a brief *depolarization* of the cell membrane, that is, the polarity of the membrane (positive outside, negative inside) reverses

b. The action potential is triggered by a threshold level stimulus
c. The stimulus causes the sodium gates to open, flooding the interior of the cell with sodium ions and reversing the polarity
d. Depolarization lasts only a fraction of a second; once the nerve impulse is transmitted, the sodium gates snap shut
e. Potassium ions pour out of the cell, following the concentration gradient; they also are repelled by the positive charge inside the membrane
f. The exit of potassium restores the resting potential

5. *Voltage-gated channels* create the action potential
   a. These channels open and close in response to the electrical gradient across the cell membrane
   b. When a stimulus begins depolarizing the membrane, the slight change in potential opens the voltage-gated sodium channels
   c. The influx of sodium ions depolarizes the membrane, opening the voltage-gated potassium channels
6. Nerve impulses are *self-propagating;* once initiated, an action potential continues automatically in one direction until it reaches the end of the nerve fiber
   a. As the nerve impulse travels along the axon, the sodium gates close behind it
   b. The sodium gates cannot reopen until the resting potential is restored; thus, the impulse cannot reverse and travel back along its original path
   c. A short delay *(refractory period)* occurs before the resting potential is reestablished and the nerve is ready to conduct another impulse

C. Transmission across a synapse

1. Nerve impulses are transmitted between neurons chemically
   a. The tip of each axon is composed of fine branches, each with a terminal swelling known as the *synaptic ending*
   b. The branched axon tip lies close to the dendrite of the adjacent neuron; this area is called the *synapse*
   c. The gap between neurons is the *synaptic cleft*
2. Transmission of a nerve impulse across a synaptic cleft is accomplished through chemicals called *neurotransmitters,* which are stored in the vesicles of the synaptic ending
   a. When a nerve impulse reaches the end of the axon, the membrane in that region becomes permeable to calcium ions
   b. Calcium ions trigger the release of neurotransmitter molecules into the synaptic cleft
   c. The neurotransmitters diffuse across the synaptic cleft and bind with cell surface receptors on the dendrite of the adjacent neuron
3. Neurotransmitters are removed quickly from the synaptic cleft
   a. In some synapses, enzymes degrade the neurotransmitter molecules
   b. In others, the neurotransmitter molecules are reabsorbed for reuse or disposal
4. Neurotransmitters can excite or inhibit impulse transmission

D. Peripheral nervous system

1. The peripheral nervous system consists of the nerve processes that connect the central nervous system to other parts of the body
2. It is divided into the somatic and the autonomic nervous systems

a. The *somatic* or *voluntary nervous system* is subject to conscious control; it responds to environmental stimuli and carries motor impulses to skeletal muscles
b. The *autonomic* or *involuntary nervous system* functions in homeostatic control
3. The autonomic nervous system is further divided into the parasympathetic and sympathetic systems
a. The *parasympathetic* division controls the digestive, circulatory, excretory, and endocrine organs; it stimulates smooth and cardiac muscle
b. The *sympathetic* division activates body systems for quick action (the *"fight or flight"* response); sympathetic fibers facilitate contractions of skeletal muscle by stimulating heart rate, metabolic rate, and energy production

E. Central nervous system

1. The central nervous system consists of the brain and the spinal cord
2. The spinal cord has two main functions
a. It carries information to and from the brain
b. It integrates simple responses to specific stimuli
(1) Integration by the spinal cord is called *reflex action*
(2) Reflexes are automatic actions that require no conscious thought
(3) Examples include the patellar reflex (knee jerk) and eye blinking; moving a hand from a hot stove also is a reflex action
3. The evolution of complex behaviors in vertebrates is correlated with an increase in brain size and complexity
4. In the complex brains of birds and mammals, the three basic brain regions—hindbrain, midbrain, and forebrain—are subdivided into additional functional regions
a. The *hindbrain,* which consists of the medulla oblongata, pons, and cerebellum, regulates homeostasis and coordinates movement
(1) The *medulla oblongata* and the *pons* are the closest regions to the spinal cord
(a) The medulla has control centers for several autonomic functions, including heart beat, breathing, digestion, and blood pressure; centers in the pons also control some of these functions
(b) The medulla has reflex centers for vomiting, coughing, swallowing, and other functions
(c) All the sensory and motor neurons connecting with higher brain regions pass through the medulla and the pons
(2) The *cerebellum* functions in muscle coordination, balance, and equilibrium
(a) It receives and processes sensory information, including the position of the body (received from the inner ear) and extension of muscles
(b) It adjusts body movement to respond to visual information (for example, in hand-eye coordination)
b. The *midbrain* is a relay and integration center for sensory information
(1) The hindbrain and the midbrain together make up the *brain stem*
(2) Brain stem functions generally are not under conscious control
c. The *forebrain* is highly developed and contains two main regions—the diencephalon and the telencephalon

d. The *diencephalon* relays information to various brain regions and contains an important center for controlling homeostatic functions; it consists of the thalamus and hypothalamus
  (1) The *thalamus* receives all sensory impulses (except those from the sense of smell) and routes them to higher brain centers
    (a) It also screens information for access to the cerebrum
    (b) It orders and prioritizes sensory input so that the cerebrum is not overwhelmed with insignificant stimuli
  (2) The *hypothalamus* is the most important homeostatic regulating site in the body
    (a) Hypothalamic regulating centers include those for body temperature, hunger, thirst, water balance, blood pressure, and sleep
    (b) The hypothalamus also is important in sexual and mating behaviors, the "fight or flight" response, and pleasurable sensations
    (c) It regulates pituitary function and therefore links the nervous and endocrine systems

e. The *telencephalon* includes the cerebral cortex and the basal ganglia (an important center of motor control), which together make up the cerebrum
  (1) The *cerebral cortex* is the largest, most complex region of the mammalian brain
    (a) This region integrates sensory input and facilitates conscious actions
    (b) It functions in learning, creative abilities, and memory
    (c) It has two bilaterally symmetrical halves, connected by a thick band of fibers (the *corpus callosum*)
  (2) The cortex contains both sensory and motor areas
    (a) Some sensory areas receive impulses from tactile, pressure, and pain receptors; others function in vision, hearing, smell, and taste
    (b) The motor areas send impulses to the skeletal muscles in the region that they control
    (c) Sensory and motor regions of the cortex control different parts of the body
    (d) The area devoted to a specific body region is positively correlated with the importance of the sensory and motor information in that region; for example, large areas are devoted to control of the hands

IV. The Endocrine System

A. General information

1. Metabolic functions are coordinated by the endocrine system and the nervous system
2. ***Hormones*** are the chemical messengers of the endocrine system
   a. They are secreted by ductless glands specialized for their production
   b. They generally travel through the circulatory system to target organs, which have receptors for specific hormone molecules

B. Hormone structure and mode of action

1. Hormones are classified according to their chemical structure with no reference to function; hormones within the same chemical class may have widely divergent functions
2. *Steroid hormones* are fat-soluble and have a ring structure derived from cholesterol
   a. They are produced by the adrenal cortex, testes, and ovaries
   b. They enter the nucleus of the target cell and activate transcription of specific genes
      (1) Some nonsteroid hormones have the same mode of action in the nucleus
      (2) Thyroxine, produced by the thyroid gland, is an example
   c. Examples of steroid hormones are cortisol and the sex hormones testosterone and estrogen
3. *Peptide hormones* are short chains of amino acids or large proteins
   a. They are produced by most endocrine glands
   b. They cannot penetrate cell membranes; their regulatory activity occurs from the cell surface
      (1) Peptide hormones bind to receptor proteins on the target cell surface
      (2) The hormone-receptor complex stimulates the cell to produce intracellular messenger molecules, which transfer the chemical signal to the inside of the cell
         (a) The intracellular messenger molecules are called *second messengers* because they carry the message from the *first messenger* (the peptide hormone)
         (b) *Cyclic adenosine monophosphate (cAMP)* and *inositol triphosphate (IP_3)* are important second messenger molecules
      (3) The intracellular messenger molecules stimulate enzyme activity and initiate a series of reactions that lead to the formation of a desired end product
   c. Peptide hormones do not affect protein synthesis; they stimulate the activity of enzymes and other proteins that already are present in the target cell

C. Types of hormones

1. Hormones are secreted by many vertebrate tissues and organs; some glands secrete more than one type of hormone
2. The *anterior pituitary* produces several different peptide hormones
   a. Four of these hormones stimulate the release of hormones from other endocrine glands
      (1) *Thyroid-stimulating hormone (TSH)* regulates the release of thyroid hormones
      (2) *Adrenocorticotropin* stimulates the production and secretion of steroid hormones by the adrenal cortex
      (3) The *gonadotropins (follicle-stimulating hormone* and *luteinizing hormone)* regulate the activity of the male and female gonads (testes and ovaries)
   b. The anterior pituitary also secretes growth hormone, prolactin, and melanocyte-stimulating hormone

(1) *Growth hormone* stimulates skeletal growth and protein synthesis; growth hormone levels determine the height (or ultimate size) of an animal

(2) *Prolactin* has various functions in the different vertebrate classes; in mammals, it stimulates mammary gland development and milk production

(3) *Melanocyte-stimulating hormone* controls the action of pigment-containing cells, causing skin color changes in some vertebrates; its function in humans is unknown

3. The *hypothalamus* regulates homeostasis, including osmotic balance, heart rate, and body temperature; it also integrates the nervous and endocrine systems
   a. Hypothalamic hormones that regulate the activity of the anterior pituitary, and thus other endocrine glands, are called *releasing factors*
   b. Releasing factors are part of a *negative feedback control system*
   c. In negative feedback control, high levels of a substance shut off production of that substance
      (1) Control of thyroxine secretion is an example of a negative feedback loop
         (a) The hypothalamus produces a releasing factor (TRH) that activates the anterior pituitary to produce TSH
         (b) TSH binds to receptor proteins on target cells in the thyroid, where it stimulates the cells to release cAMP
         (c) Inside the cells, cAMP activates the synthesis and release of thyroid hormones, such as thyroxine
      (2) The feedback loop is controlled at three points
         (a) High levels of TSH inhibit TRH secretion by the hypothalamus
         (b) High levels of thyroxine inhibit TSH secretion by the anterior pituitary
         (c) High levels of thyroxine also inhibit TRH secretion by the hypothalamus
4. The *thyroid gland* produces *thyroxine,* which regulates metabolic rate and development; another thyroid hormone, *calcitonin,* lowers blood calcium levels
5. The *parathyroid glands* secrete *parathyroid hormone,* which raises blood calcium levels
6. The *adrenal glands,* located near the kidneys, contain two functional subunits; both secrete hormones in response to environmental stress
   a. The *medulla* (inner region), which is controlled by the autonomic nervous system, secretes two hormones that initiate the "fight or flight" response
      (1) *Epinephrine* (also called *adrenaline*) and *norepinephrine* (also called *noradrenaline*) stimulate energy-providing metabolic reactions and increase heart rate and blood pressure
      (2) They also function as neurotransmitters
   b. The *cortex* (outer region) secretes *corticosteroids*
      (1) *Glucocorticoids,* such as *cortisol,* stimulate glucose metabolism to provide an energy boost
      (2) *Mineralocorticoids,* such as *aldosterone,* function in osmoregulation
7. The *pancreas* secretes two hormones that control glucose metabolism
   a. *Insulin* is secreted when blood glucose levels are high; it stimulates liver, adipose, and muscle cells to metabolize glucose and store the excess as glycogen or fat

b. *Glucagon* is secreted when blood glucose levels are low; it stimulates liver cells to convert glycogen back to glucose and release it into the bloodstream

8. The *gonads* produce hormones of the reproductive system, such as *estrogen* and *testosterone*
9. The *pineal gland,* located in the brain, secretes *melatonin*
   a. This gland has light-sensitive cells that respond to changes in the photoperiod and regulate biologic clock mechanisms (daily and seasonal activity cycles)
   b. It plays a role in the timing of vertebrate reproductive cycles
10. Several nonendocrine glands, such as the heart, kidney, and digestive organs, also secrete hormones

V. The Immune System

A. General information

1. Vertebrates have complex defense mechanisms against entry of pathogens (disease-causing agents), such as bacteria, viruses, fungi, and protozoans
2. *Nonspecific defenses* prevent substances from gaining entry to the body
3. If nonspecific defenses fail, *specific defenses* destroy invaders in the tissues or interstitial fluids

B. Nonspecific defenses

1. Nonspecific defenses are physical and chemical barriers against invasion
   a. The skin is the primary organ of passive defense
   b. Internally, mucus membranes lining the digestive and respiratory tracts trap foreign particles
   c. The acidic pH of gastric fluids in the stomach inhibits the growth of most types of microorganisms
   d. Colonies of symbiotic bacteria inhabit all surfaces that are in contact with external substances and prevent encroachment by foreign species; such colonies are located on the skin surface, and in the esophagus, lungs, intestine, and vagina
2. Additional nonspecific defenses come into play when microbes gain entry to the body
   a. Phagocytic white blood cells engulf and destroy microorganisms
      (1) Neutrophils, eosinophils, and monocytes are phagocytic cells
      (2) Monocytes develop into macrophages, the main phagocytic cells of the immune system
   b. Phagocytic cells can move throughout the body
   c. They also are located in various organs, such as the spleen, lymph nodes, liver, kidneys, lungs, and joints
3. Specialized plasma proteins also play a role in nonspecific defenses
   a. These cells, called *complement proteins* (part of the complement system), circulate in the bloodstream and are activated by contact with a foreign invader
   b. Complement proteins may create chemical gradients that attract phagocytic cells to the infection site

c. They may secrete a substance onto the invader's cell surface that attracts phagocytic cells

d. They may attack the cell wall or cell membrane of invaders and dissolve it

4. Inflammation is the result of a localized immune system response, in which invaders are destroyed and damaged tissues are repaired

C. Specific defenses

1. Specific defenses use recognition proteins on the cell surface to distinguish between the body's cells and those of foreign invaders
   a. Nonself cells and chemicals that elicit an immune system response are called *antigens*
   b. Specific antigens are targeted for attack by the immune system
2. The two main defensive branches of the immune system are the antibody-mediated immune response and the cell-mediated immune response
3. The *antibody-mediated immune response* (also called the *humoral immune response*) produces antibodies
   a. *Antibodies* are receptor molecules with binding sites for a specific type of antigen; they circulate freely or are bound to cell membranes
   b. Antibodies are derived from *B lymphocytes* (also called *B cells*)
   c. When a B cell encounters an antigen, it becomes activated
      (1) Activated B cells produce many *plasma cells* (called *clones*), which secrete tailored antibodies against the antigen
      (2) To be activated, a B cell must be stimulated by a *helper T cell*
   d. In most cases, the antibodies and antigens form a large clump, called the *antigen-antibody complex*
      (1) The antigen-antibody complex attracts the attention of other immune system components
      (2) The complex may be phagocytized by neutrophils or macrophages
      (3) It also may activate the complement system to destroy the invader or tag the pathogen to attract phagocytic cells
   e. Members of the clone that do not participate in antibody production become *memory B cells*
      (1) Memory cells circulate in the bloodstream
      (2) They respond to future invasions by the same antigen by rapidly producing tailored plasma cells
4. The *cell-mediated immune response* destroys pathogens that have already entered host cells; many pathogens, such as viruses, are intracellular parasites that can reproduce only inside a host cell
   a. This immune response involves the action of *T cells*
      (1) T cells are produced in the bone marrow but mature in the thymus
      (2) They respond only to foreign antigens on the surface of the body's own cells; the antigens combine with the body cell's own recognition proteins to form a *major histocompatibility complex (MHC)*
   b. The cell-mediated immune response begins when body cells engulf a foreign invader and attach fragments of the foreign antigens to the MHC proteins on their cell membrane
      (1) These cells are known as *antigen-presenting cells (APC)*
      (2) Macrophages and B cells commonly become APCs, but any body cell can perform the same function if invaded by a pathogen

c. The MHC-antigen complex activates the T cells, directing them to locate and destroy all cells displaying the same antigen

5. Several types of T cells play important roles in the cell-mediated immune response
   a. *Cytotoxic* (or *killer*) *T cells* attack cells bearing a specific foreign antigen
   b. *Helper T cells* strengthen the response of other immune cells (in both humoral and cell-mediated immune responses)
   c. *Suppressor T cells* release cytokines that inhibit the activity of other T cells and B cells
      (1) When the number of suppressor T cells reaches a threshold level, the immune response is terminated
      (2) Suppressor T cells probably are activated at the end of the immune cycle, when the antigen has been destroyed and further immune response is not necessary
   d. *Memory T cells* are similar in action to memory B cells
      (1) A small population of these cells persists, tuned to the specific antigen that triggered the immune response
      (2) They provide a rapid response capability to future invasions by the same pathogen

Study Activities

1. Describe the structure and function of the circulatory system.
2. Describe mechanisms of temperature control.
3. Describe vertebrate methods of excretion and osmoregulation.
4. Describe the structure and function of the nervous system.
5. Describe the structure and function of the endocrine system.
6. Describe the structure and function of the immune system.

17

Animal Behavior

Objectives

After studying this chapter, the reader should be able to:
- Understand the mechanisms underlying animal behavior.
- Differentiate between proximate and ultimate causes of behavior.
- Differentiate between innate and learned behaviors.
- Understand how natural selection acts in the evolution of behavior.
- Understand how the basic principles of optimality theory apply to feeding, reproductive, and social behaviors.

I. The Study of Animal Behavior

A. General information

1. The study of animal behavior, also called *ethology,* involves the observation of an animal's responses to environmental stimuli
2. Both simple and complex behavior patterns are analyzed to discover how and why animals behave as they do
3. The objective of most animal behaviorists is to describe how animals behave in their native habitats
   a. Many animal behavior studies are conducted in the laboratory
   b. These results usually are compared with observations of natural behaviors in the wild

B. Questions about animal behavior

1. "How" questions examine the *proximate* (immediate) causes of behavior
   a. Proximate questions investigate the physical mechanisms that make a specific behavior possible
      (1) The anatomic and physiologic capabilities of a species are studied to answer these questions
      (2) A specific behavior may be analyzed, for example, in terms of the sensory receptors involved in receiving the stimulus, transmission of nerve impulses, integration of sensory input in the brain, and ability of muscles to respond to the stimulus
   b. Proximate explanations of behavioral traits are related to the genetic makeup, embryologic development, physiology, or psychology of the species in question
2. "Why" questions examine the *ultimate* (evolutionary) causes of behavior

a. Questions about ultimate causes are approached in terms of the survival benefits that certain behavior patterns confer on the individual and the species
b. Ultimate explanations of behavior are related to the reproductive or functional value of a behavioral trait and the evolutionary history of the trait's development
c. Ultimate explanations of behavior often are based on Darwin's principles of natural selection (see Chapter 2, Evolution of Animal Diversity)

II. Principles of Ethology

A. General information

1. Behaviors are either learned or innate (instinctive)
   a. *Innate behaviors* are present at birth; experience or learning is not required to accurately perform these behaviors
   b. *Learned behaviors* are modified and refined through life experiences; the accuracy of these behaviors improves as the animal gains information
   c. The dividing line between innate and learned behaviors is not fixed
   d. Many behaviors have an instinctive component, but also may be modified by learning
2. Behavior is *heritable*
   a. The genes contain instructions for the development of anatomic and physiologic mechanisms that receive and process sensory information
   b. Because behavioral traits have a genetic basis, they are susceptible to changes through natural selection

B. Innate behaviors

1. Innate behaviors have a set sequence of events
   a. The sequence and nature of the events does not change with experience
   b. Each member of the species performs the sequence in the same way
2. Innate behaviors, once triggered, usually are carried out to completion; this is called a *fixed action pattern (FAP)*
   a. Once initiated, an FAP is difficult to stop short of completion
   b. For example, the swallow reflex is a human FAP; once a mouthful of food has started to be swallowed, the process cannot be halted halfway
   c. Each step is triggered by the preceding step, through to completion
3. The FAP is triggered by a simple sensory cue, called a *releaser*
4. FAPs are common in humans and other animals
   a. A typical FAP is the yawn
      (1) The sequence of actions in a yawn is relatively constant from person to person
      (2) Most yawns last about 6 seconds
      (3) Once a yawn begins, it is difficult to stop short of completion
      (4) Seeing or hearing someone yawn stimulates yawning in the observer (yawning is a *releaser* of yawning in other humans)
   b. Young herring gulls have a fixed sequence for begging behavior
      (1) The sequence is triggered by the red spot on the parent gull's bill
      (2) Any long, thin, bill-like object with a contrasting dot at the end will induce this behavior

C. Learned behaviors

1. Learned behaviors are modified based on individual experiences; these behaviors range from simple to complex
2. The ability to modify behaviors based on new information can increase an animal's ability to survive and reproduce
    a. It also can help individuals adjust to variable environmental conditions that could not be anticipated before birth
    b. Examples of such conditions include the location and nature of prey species, changing prey availability, and local language and customs
3. There are several categories of learned behaviors
    a. In *classical conditioning,* an animal learns to associate an automatic (instinctive) response with a new stimulus (one that does not normally trigger the response)
        (1) The concept of classical conditioning was developed by Ivan Pavlov in his experiments with dogs
        (2) Dogs were conditioned to salivate (the normal response to the presence of food) at the ring of a bell, even when food was absent
    b. In *operant conditioning,* an animal learns to associate a voluntary activity with the results that follow
        (1) Toads or birds that attempt to eat bad-tasting insects learn to avoid those species in the future
        (2) Rats can learn that pressing a bar in their cage releases a food pellet, thus reinforcing the bar-pressing behavior
        (3) Humans can be taught to regulate physiologic processes (such as blood pressure) through operant conditioning
    c. In *habituation,* animals learn to ignore a repeated stimulus, if the stimulus has no important positive or negative consequences
        (1) Urban birds on telephone wires do not fly away from people or automobile traffic below
        (2) Humans become accustomed to background noises and eventually do not perceive them
    d. *Imprinting* is a time-sensitive form of learned behavior
        (1) The concept of imprinting was developed by Konrad Lorenz
        (2) Young animals, especially birds, learn to form an association with a particular individual by the presence of certain key stimuli
        (3) This association, which is permanent, can be formed only during a specified period of development; in birds, this usually occurs upon hatching
            (a) Young geese imprinted on Lorenz when hatching; they followed him everywhere, as they would their natural mother
            (b) As adults, the imprinted animals direct their sexual behavior toward members of whatever species they had imprinted on in their youth; Lorenz's geese, for example, attempted to court humans
        (4) Apes (such as gorillas and orangutans) in zoos sometimes are raised by their human keepers
            (a) As infants, they imprint on humans instead of other members of their species
            (b) As adults, they also inappropriately direct their courting behavior toward humans

(5) Some species of birds, such as the white crowned sparrow, learn their songs through the imprinting process
   (a) Nestlings imprint the songs of adult birds during a specific period of development; in the absence of this stimuli, their song will be abnormal
   (b) The birds can learn any song during the critical imprinting period, even those of other species

e. *Spatial learning* involves mapping and orienting to environmental landmarks
   (1) Female digger wasps build their nest and then leave to forage for prey; regardless of the direction of their foraging trip, they can return to the tiny, hidden nest entrance
   (2) Birds, such as blue jays or chickadees, that store food can remember and retrieve it from hundreds of locations

f. *Insight learning* encompasses the higher reasoning skills
   (1) When an animal encounters a new situation, it develops a response based on past experiences
   (2) Through reasoning, an animal can respond to a problem-solving situation without trying all possible solutions
   (3) This ability varies with the species; it is well developed in primates

III. Adaptive Value of Behavior

A. General information

1. Evolutionary mechanisms lead to the development of specific behavioral patterns
2. Because behavioral traits have a genetic basis, they are susceptible to changes through natural selection
3. Behaviors that improve reproductive success (the ability to produce more offspring that reach reproductive age) become more prevalent within a population
4. *Optimality theory* uses the principles of natural selection to make specific predictions about the optimal solution to a problem; optimal solutions are those that increase an individual's reproductive success more than other solutions

B. Feeding behavior

1. Locating and acquiring food resources is an energy-intensive behavior; each species has its own diet and characteristic foraging methods
2. The *optimal foraging theory* states that, through natural selection, feeding behaviors are modified and optimized within a species so that an individual can acquire maximal food resources with minimal risk and energy output
   a. This theory predicts that an animal that optimizes its feeding strategy will spend less time and energy foraging for food
   b. The animal, therefore, has more time and energy available for producing offspring, thus maximizing its reproductive success
   c. The optimizing animal will produce more offspring for the next generation, while animals that forage less efficiently will contribute fewer offspring
   d. Because some components of behavior are hereditary, optimizing behaviors become firmly fixed and increase in frequency in subsequent generations
3. Optimality theory has been applied to herbivores, carnivores, and omnivores
   a. Optimal foragers should preferentially select larger, higher-calorie diet choices

(1) Large prey require almost the same energy input to locate and consume as small prey
(2) By focusing attention on larger prey, the animal maximizes its cost/benefit ratio; it receives the most calorie benefits for its foraging costs

b. Optimality theories have been tested on many species of vertebrates; results show that adherence to optimality predictions is not universal within the animal kingdom

4. Factors unrelated to the prey caloric content may prevent animals from feeding with maximum efficiency
   a. Foragers are themselves subject to predation; minimizing their exposure and risk can be an important consideration in prey selection
   b. Competition for limited food resources can influence food selection, leading animals to choose more abundant or easily attained food
   c. Nutritional constraints also can influence food choices; relatively low-calorie prey may be selected because they contain key nutrients required in the diet

C. Mating behavior

1. Males and females differ in their tactical approach to reproductive success
   a. Males make many small gametes (sperm)
      (1) Each gamete represents a relatively low energy investment
      (2) In most species, males try to fertilize as many eggs as possible, and provide little or no parental care
   b. Females make fewer, large gametes (eggs)
      (1) Each gamete represents a relatively high energy investment
      (2) Females often invest additional energy in parental care
2. Sexually reproducing species usually do not mate randomly; competition for the opportunity to mate is the rule, rather than the exception
   a. Each individual tries to maximize its genetic contribution to the next generation
   b. In general, the sex with the highest parental investment in the next generation (the female) selects the mate to father her offspring
   c. Competition for mates can lead to *sexual selection,* an evolutionary change in the physical characteristics and reproductive behavior of males and females within a species
      (1) Sexual selection has resulted in the appearance of male characteristics such as the peacock's large, colorful tail; lion's mane; and deer antlers
      (2) It also has resulted in behavioral characteristics such as elaborate courtship rituals, defense of feeding or nesting territories, and defense of a harem by dominant males

D. Social behavior

1. Social living has costs and benefits, which can be measured in terms of reproductive success
2. Not all species exhibit social behavior; in many circumstances, solitary individuals have greater reproductive success than those congregating in groups
3. Social behavior has several advantages
   a. Social groupings can enhance predator defense

(1) Members of a community may signal the group when a predator is sighted (as with prairie dogs) or cooperate to repel a predator (as with gulls)

(2) Predators can be distracted or confused by a large mass of prey (such as a school of fish), allowing most individuals to escape

b. Social groupings can enhance predation efficiency

(1) Members of a group (wolves or hyenas, for example) may cooperate to capture large prey

(2) Members of a group (such as chimpanzees) may cooperate to stalk a prey animal and chase it into the open

c. Social groupings can facilitate the defense of limited resources against other members of the same species (as lion prides defend hunting territories)

d. Social groupings can provide improved infant care and protection (as with colonial nesting birds and many primates)

e. Advantageous learned behaviors can spread quickly through the group (such as consumption of novel foods)

f. Social groupings provide opportunities for division of labor (as with social insects such as ants, bees, and termites)

4. Social behavior also has disadvantages

a. Crowding in a limited area increases competition for food, living space, mates, nest sites, nest materials, and other limited resources

b. Social groupings increase the risk of transmitting infectious diseases and parasites

c. Cooperative parental care may be exploited by some individuals, who leave others to care for their offspring

d. Offspring may be deliberately or accidentally killed by other members of the group (elephant seal pups often are crushed to death by the huge males)

5. Animals form social groupings only when the benefits of group living outweigh the costs

Study Activities

1. Differentiate between proximate and ultimate causes of behavior.
2. Differentiate between innate and learned behaviors.
3. Compare and contrast four types of learned behaviors.
4. Write a short essay that discusses how natural selection acts to modify behavior.
5. Discuss how the basic principles of optimality theory relate to feeding, reproductive, and social behaviors.

18

Animal Ecology

Objectives

After studying this chapter, the reader should be able to:
- Become familiar with the various levels of ecologic studies.
- Understand how abiotic and biotic factors affect community structure.
- Describe factors that determine biodiversity.
- Explain the niche concept.

I. The Study of Ecology

A. General information

1. Different factors in an animal's environment have an impact on the animal's ability to survive and reproduce
   a. *Abiotic* factors are the nonliving components of the environment, such as temperature, light, and water
   b. *Biotic* factors are other living organisms in the environment
2. Animals interact with their environment
   a. They are affected by environmental conditions
   b. In turn, their presence and activities produce important changes in their habitat
3. Ecologists study various aspects of animal/environment interactions
   a. *Physiologic ecology* examines the structural and functional ways in which individual animals meet the challenges of their environment
   b. *Population ecology* studies a specific group within a species, focusing primarily on factors that lead to changes in population size
   c. *Community ecology* looks at interactions among all the animals that inhabit a defined area, focusing on such factors as predation, competition, and symbiotic associations
   d. *Ecosystem ecology* incorporates abiotic factors into a community study; it addresses areas such as energy flow and cycling of chemicals through the biotic and abiotic components of the ecosystem

B. Important abiotic factors

1. Abiotic factors can have a strong impact on the distribution and success of animals within an environment
2. *Temperature* is a critical factor in animal distribution; most species cannot maintain body temperatures more than a few degrees above or below the ambient temperature

3. *Water* also plays a role in determining animal distribution; a relative abundance of organisms in various habitats generally is correlated with the availability of water
   a. Aquatic organisms must cope with osmotic differences between their body fluids and the surrounding medium
   b. Terrestrial organisms must prevent desiccation and have adaptations to reduce water loss
4. *Light* is important in the lives of plants and animals
   a. Plants require solar energy to perform photosynthesis
   b. Many animals have life cycles that are attuned to the photoperiod; they may be nocturnal, diurnal, or crepuscular (active at dawn and dusk)
   c. Reproduction and migration often depend on photoperiod cues, such as changing day length
5. *Soil* type has an influence on plant communities, and thus on the distribution of herbivores; animals also require various plants for shelter and building materials
6. *Wind* amplifies the effects of environmental temperature
   a. It contributes to desiccation by increasing the rate of evaporation
   b. It also increases an animal's rate of heat loss to the environment, by evaporation and convection

II. Biodiversity

A. General information

1. Communities vary greatly in species composition; some communities have few species, others have many species
2. *Biodiversity* is the measure of the number of different species in a community or ecosystem

B. Factors affecting biodiversity

1. Species diversity usually is negatively correlated with the population density of individuals within a species
   a. Tropical rain forests have extremely high biodiversity, but the number of individuals within each species is low
   b. In habitats with low biodiversity, such as tidal mud flats, the biomass of each species is remarkably high
2. In general, biodiversity decreases as latitude increases (in both terrestrial and marine ecosystems)
   a. This effect may be due to the stability of ecosystems
   b. Ecosystems at high latitudes undergo great seasonal changes in climate
3. Biodiversity also decreases with increasing altitude (probably because of stressful environmental conditions)
4. Ecologic theory predicts that animals inhabiting stressful, variable regions will tend to be r-selected species, while those in stable habitats with less seasonal stress will be primarily k-selected species
   a. *R-selected species* reproduce rapidly, have many young of which few survive to adulthood, and have short life spans; rodents are a typical example
   b. *K-selected species* produce few offspring with better chances for survival and have long life spans; examples include the larger birds and mammals

5. Northern latitudes have primarily r-selected species (such as rodents), but several k-selected species (such as polar bears, seals, and whales) also are present; the reverse is true for tropical ecosystems

C. The niche concept

1. Each animal species has a *niche*, or well-defined life-style
    a. The niche includes all of the interactions between a species, its physical environment, and the other biotic components of the habitat
    b. Niche factors include the living space required, food choices, preferred temperature and moisture levels, and activity periods
2. The niche concept predicts that, unless all resources are overabundant, no two species can coexist in exactly the same niche in any community
    a. If a resource is limited in supply, one species will compete more efficiently for that resource
    b. As a result of this competition, the less efficient competitor is excluded
    c. The excluded species may die out or begin to use slightly different resources, thereby defining a new niche
3. Ecologic theory predicts that, in more stressful habitats, resident species will have a high tolerance for variable environmental conditions; in stable habitats, they will have a low tolerance for environmental change
    a. Species with narrow tolerance ranges have narrow, rigidly defined niches
    b. Tropical forest communities, for example, are composed of many tightly distributed species, each occupying a small niche
4. When species coexist within a habitat, several species may exploit the same resource in slightly different ways; this approach is known as *resource partitioning*
    a. Species may use the same space but have different activity periods (nocturnal versus diurnal)
    b. Species may partition food resources
        (1) On the African plains, some grazers feed on the upper parts of grasses, others on the stems
        (2) As grazers trample the tall grasses, they expose low-growing herbs, which are fed upon by still other species
        (3) Rodents and insects feed on vegetation scraps and undigested plant material in the feces of herbivores

Study Activities

1. Discuss the various levels of ecologic studies.
2. Explain how abiotic and biotic factors affect community structure.
3. Write a short essay on the factors that affect ecosystem biodiversity.
4. Explain the niche concept.

Appendix

Selected References

Index

Glossary

Aboral—end opposite the mouth in a radially symmetrical organism
Acoelomate—having a solid body, with no body cavity surrounding the gut
Adenosine triphosphate (ATP)—important energy storage molecule in all aerobic organisms
Aerobic—requiring oxygen for metabolic functions
Allele—alternate form of a gene that is expressed as various phenotypes
Allometric growth—different rates of growth of body parts during embryologic development
Alternation of generations—basic cnidarian life cycle in which an asexual polyp form alternates with a sexual medusa form
Ammonia—nitrogenous waste product of aquatic organisms
Amoeboid movement—movement by extrusion of pseudopodia; found in single cells
Anaerobic—not requiring oxygen for metabolic functions
Analogous—characteristics that are similar in function but not in evolutionary or embryonic origin; arise through convergent evolution
Aquatic—living exclusively in water
Asymmetrical—no defined body form
Autotroph—organism that can synthesize organic materials from inorganic elements using light or chemical energy
Bilateral symmetry—having a distinct head end and a body plan in which the right and left sides are mirror-image halves
Binomial nomenclature—two-part system for naming species
Bioluminescence—method of light production in animals that involves the breakdown of specialized proteins
Biradial symmetry—modified form of radial symmetry in which one body part is single or paired; only two planes passing through the central axis yield mirror-image halves
Blastocoel—fluid-filled cavity within the blastula, an early stage of embryonic development
Blastopore—external opening in the gastrula (invagination) stage of embryonic development that may develop into the mouth or anus
Budding—method of asexual reproduction similar to binary fission, but the cytoplasm is divided unequally, and the daughter cell usually is much smaller than the parent
Camera eye—type of eye typical of vertebrates and cephalopod molluscs; light impacts photoreceptors in the retina of the eye
Catastrophism—mechanism of evolutionary change proposed by Cuvier; mass extinctions in the fossil record are caused by a series of catastrophic events with subsequent repopulation by migration or new creation events
Cecum—pouch-like extension of the stomach or intestinal wall
Cephalization—concentration of nerve cells and sense organs in the head area
Cerebral ganglia—aggregation of nerve cells at the anterior end of the body in some cephalized animals; functions as a simple brain
Character—observable trait or feature used to separate organisms and assign them to different taxonomic groups
Chemoreceptor—type of sensory receptor that is sensitive to chemicals
Chitin—nitrogenous polysaccharide that is an important component of exoskeletons
Chlorocruorin—iron-containing respiratory pigment; found in worms
Chromatophore—pigment-containing cells that allow epithelial color change
Chromosome—condensed strand of DNA
Cilium—hair-like organelle on the surface of some eukaryotic cells; used for feeding and locomotion in protozoans and for moving materials across the cell surface in metazoans

Cladistics—classification system based solely on evolutionary history and degree of common descent
Cladogram—branching diagram that displays a cladistic classification
Cleavage—mitotic division of the zygote
Closed circulatory system—type of circulatory system in which blood is confined to vessels or vessel-like channels as it circulates through the body
Collagen—chief structural protein of invertebrates
Colony—association of unicellular or multicellular organisms of the same species; division of labor may occur among the individual members of the colony
Commensalism—form of symbiotic association that benefits one participant while the other neither benefits nor is harmed
Complex eye—see *camera eye*
Compound eye—type of eye common in arthropods; composed of a collection of separate units, each having its own field of vision and its own nerve
Convergent evolution—process by which unrelated organisms independently develop similar characteristics through adaptation for similar environments
Cryptobiosis—resistant, dormant stage in the life cycles of several pseudocoelomate phyla
Cuticle—noncellular protective layer secreted by the epidermis of pseudocoelomates
Deuterostome—animals that have a type of embryonic development characterized by radial cleavage, with the mouth arising anterior to the blastopore
Dioecious—male and female sexes in separate individuals
Diploblastic—body tissues that are derived from only two of the three embryonic germ (basic tissue) layers: endoderm and ectoderm; mesoderm is lacking
Ectoderm—outermost portion of the three embryonic germ layers; gives rise to the epidermis and nervous system
Ectotherm—see *poikilotherm*
Endoderm—innermost portion of the three embryonic germ layers; gives rise to the epithelial tissue lining the digestive tract and digestive organs
Endoparasite—parasitic organisms that live within the bodies of animal hosts
Endotherm—see *homeotherm*
Eucoelomate—having a true coelom, a body cavity derived from and lined by mesoderm
Eukaryotic—cells with organelles surrounded by membranes; typical of protozoans and metazoans, excluding bacteria
Eumetazoa—collective name for all animals at the tissue level of organization and beyond
Euryhaline—animals that can survive a wide range of salinity changes in their habitat
Evolution—change in the frequency of a trait in a population
Evolutionary taxonomy—classification system that separates species into groups based on the number of shared homologous characters and evolutionary history
Flagellum—filamentous organelle used for locomotion in protozoans and other eukaryotic cells
Fragmentation—form of asexual reproduction in which the body splits into segments and each part regenerates a new individual; commonly seen in annelid worms
Gastrovascular cavity—central body cavity with just one opening that functions in digestion, respiration, and circulation; found in phylums Cnidaria and Ctenophora
Gastrula—stage of embryonic development during which all three germ layers form
Gene pool—all of the alleles possessed by all individuals in a population
Genetic drift—accidental changes in gene frequencies that occur when a small number of individuals become isolated from the main population
Genotype—total genetic or hereditary makeup of an organism
Georeceptor—organ of equilibrium and balance

Germ layers—three basic embryonic tissue layers of eumetazoans; all organs and body structures develop from these layers
Hemerythrin—iron-containing respiratory pigment; found in worms
Hemocoel—body cavity in animals that have an open circulatory system
Hemocyanin—copper-containing respiratory pigment; found in molluscs and crustaceans
Hemoglobin—iron-containing respiratory pigment; found in almost all vertebrates and many invertebrates
Hemolymph—combination of blood and coelomic fluid found in open circulatory systems
Hermaphrodite—both sexes present in the same individual; in most cases, eggs and sperm are not produced at the same time, thus avoiding self-fertilization
Heterotroph—organisms that must obtain organic raw materials (food) from the environment; they cannot perform photosynthesis
Hibernation—prolonged dormant period in which animals live on stored fat reserves
Homeostasis—maintenance of a physiologic steady state
Homeotherm—endothermic animal that uses metabolic heat to maintain a stable body temperature
Homologous—characters that are inherited from a common ancestor and show a similar pattern of embryonic development
Hormone—chemical messenger of the endocrine system
Hydrostatic skeleton—skeletal support in which a fluid-filled body cavity is pressurized by a muscular body wall
Hypertonic—a solution with a higher osmotic (solute) concentration than some reference solution; when separated by a selectively permeable membrane, the hypertonic solution will gain water from the reference solution
Hypotonic—a solution with a lower osmotic (solute) concentration than some reference solution; when separated by a selectively permeable membrane, the hypotonic solution will lose water to the reference solution
Inheritance of acquired characteristics—mechanism of evolutionary change proposed by Lamarck; organisms accumulate adaptations through constant striving to adjust to their environments
Integument—outer body covering; includes the skin and all structures associated with the skin
Intermediate host—one that houses an immature stage in the life cycle of an endoparasite
Interstitial fluids—coelomic fluids surrounding the tissues and organs
Interstitial spaces—cryptic habitat between particles of terrestrial or aquatic sediments; 22 of the 40 animal phyla inhabit interstitial spaces
Isotonic—a solution with the same osmotic (solute) concentration as the reference solution; when separated by a selectively permeable membrane, the isotonic solution experiences no net water gain or loss
Macroevolution—evolutionary change above the species level
Marine—living exclusively in sea water
Marsupial—mammals in which embryonic development is completed in pouches
Mesoderm—central portion of the three embryonic germ layers; gives rise to the skeleton, muscles, circulatory system, and many other organs
Metamere—see *somite*
Metamerism—repetitive linear sequence of similar body parts; may include both internal and external body structures; found only in phylums Annelida, Arthropoda, and Chordata
Mode—mechanism of evolutionary change
Molt—shedding and replacement of an exoskeleton; necessary to permit growth
Morphology—internal and external anatomic structure
Mutualism—form of symbiotic association that benefits both participants

Myogenic—type of muscle contraction that originates within the muscle cells themselves
Natural selection—mechanism of evolutionary change proposed by Darwin; also known as survival of the fittest
Neoteny—retention of embryonic or juvenile characteristics by the adult
Nerve net—loosely organized nervous system of interconnected nerve cells; found in phylums Cnidaria, Ctenophora, and Echinodermata
Neurogenic—type of muscle contraction that is initiated by nerve action
Neuron—functional unit of the nervous system; also called the *nerve cell*
Neurotoxin—poison that inhibits the nervous system
Nocturnal—animals that are active primarily at night
Notochord—supportive longitudinal rod found in nonvertebrate chordates
Numerical taxonomy—see *phenetics*
Open circulatory system—type of circulatory system in which blood is pumped through the heart, through arteries, and into a series of chambers that open into the body cavity
Oral—mouth end of a radially symmetrical organism
Organ—specialized centers of function, composed of more than one tissue type, in the bodies of multicellular animals
Osmoconformer—invertebrate animal that cannot regulate the osmotic concentration of its body fluids, which therefore varies with environmental salinity
Osmoregulation—maintenance of an appropriate solute and water concentration in internal body fluids
Osmoregulator—animal that can maintain the osmotic concentration of its body fluids despite fluctuations in the environment
Paedomorphosis—retention of juvenile characteristics in the adult
Paleontology—science of the study of fossils
Parasitism—form of symbiotic association that benefits one participant at the expense of the other
Parthenogenesis—form of sexual reproduction in which females produce diploid, unfertilized eggs that develop into other females only
Pedal laceration—form of asexual reproduction in phylums Cnidaria and Ctenophora; small pieces of the body break off and each develops into a new individual
Phenetics—classification system that groups organisms based on the percentage of similar characters
Phenogram—branching diagram that displays a phenetic classification
Phenotype—observable set of physical characteristics
Pheromone—species-specific communication chemicals that comprise a complex chemical language; common in insects
Phylogenetic systematics—see *cladistics*
Phylogenetic tree—branching diagram that displays an evolutionary taxonomy classification
Photoreceptor—type of sensory receptor that contains light-sensitive pigments
Placental—mammals in which embryologic development is completed in the uterus
Plankton—assemblage of free-floating plants and animals with only limited swimming capabilities
Poikilotherm—animal that has only limited metabolic control over its body temperature
Polymorphism—having two or more basic body types or sets of physical characteristics within a species
Prokaryotic—cells with organelles not surrounded by membranes; typical of bacteria
Protostome—animals that have a type of embryonic development characterized by spiral cleavage, with the mouth arising from the blastopore
Pseudocoelomate—having a body cavity derived from the embryonic blastocoel

Punctuated equilibrium—pattern of evolutionary change proposed by Eldredge and Gould; short periods of rapid evolutionary change are followed by long periods of stasis
Radial cleavage—pattern of embryonic development in which the mitotic divisions of the zygote are distributed evenly; the zygote can be divided along any plane to form two mirror-image halves
Radial symmetry—similar body parts arranged around a central axis; more than one plane through this axis yields mirror-image halves
Recognition protein—species-specific protein on the cell membrane that allows self-identification
Respiratory pigment—carrier molecule that transports oxygen in the blood; examples are hemoglobin and hemocyanin
Rheoreceptor—type of sensory receptor that senses water currents
Septum—sheet-like dividers that extend from the body wall of invertebrates
Segmentation—see *metamerism*
Somite—having body segments in a metameric sequence
Speciation—formation of new species
Species—group of organisms that share similar characteristics and interbreed freely to produce fertile offspring
Spherical symmetry—having a ball-shaped body; similar body parts are arranged around a central point; an infinite number of planes through this point yields mirror-image halves
Spiral cleavage—pattern of embryonic development in which the mitotic divisions of the zygote occur in a spiral pattern
Statocyst—type of georeceptor
Stenohaline—aquatic animals that are restricted in habitat to a narrow range of salinities
Symbiosis—close association between two organisms of different species; the association may be commensal, mutualistic, or parasitic
Systematics—study of the diversity of living organisms and the relationships among them; includes taxonomy
Tactile receptor—type of sensory receptor that is sensitive to touch and vibration
Taxonomy—identification and classification of species
Tempo—rate at which evolutionary change occurs
Thermoreceptor—type of sensory receptor capable of directly sensing differences in environmental temperature
Tissue—layers of cells with similar structure and function
Torpor—short, dormant period common in small birds and small mammals; body temperature is decreased as an energy conservation measure
Triploblastic—body tissues that are derived from all three embryonic germ (basic tissue) layers: endoderm, mesoderm, and ectoderm
Urea—water-conserving nitrogenous waste product of terrestrial invertebrates and vertebrates
Uric acid—insoluble nitrogenous waste product excreted by birds, reptiles, and insects
Vector—transmitter of pathogens
Zooxanthella—unicellular alga that lives within the tissues of many marine invertebrates in a symbiotic relationship
Zygote—cell formed by fusion of ovum and sperm; fertilized egg

Selected References

Alcock, J. *Animal Behavior.* Sunderland, Mass.: Sinauer Associates, 1988.

Brusca, R.C., and Brusca, G.J. *Invertebrates.* Sunderland, Mass.: Sinauer Associates, 1990.

Campbell, N.A. *Biology.* Menlo Park, Calif.: Benjamin/Cummings Publishing Co., 1990.

Dorit, R.L., Walker, W.F., Jr., and Barnes, R.D. *Zoology.* Philadelphia: Saunders College Publishing, 1991.

Hickman, C.P., Roberts, L.S., and Hickman, F.M. *Biology of Animals,* 5th ed. St. Louis: Mosby, Inc., 1990.

Mader, S. *Human Biology.* Dubuque, Iowa: Wm. C. Brown Publishers, 1992.

Miller, S.A., and Harley, J.P. *Zoology.* Dubuque, Iowa: Wm. C. Brown Publishers, 1992.

Moyle, P.B., and Cech, J.J., Jr. *Fishes.* Englewood Cliffs, N.J.: Prentice Hall, 1988.

Postlethwait, J.H., Hopson, J.L., and Veres, R.C. *Biology.* New York City: McGraw-Hill Publishing Co., 1991.

Index

i refers to an illustration

i refers to an illustration